MONOGRAPH

ON

Flavoring Extracts

WITH

Essences, Syrups,

AND

Colorings.

ALSO

FORMULAS FOR THEIR PREPARATION.

WITH APPENDIX.

Intended for the Use of Druggists.

By JOSEPH HARROP, Ph. G.

For some years engaged in their exclusive manufacture.

Fredonia Books
Amsterdam, The Netherlands

Monograph on Flavoring Extracts:
With Essences, Syrups, and Colorings

by
Joseph Harrop

ISBN: 1-4101-0501-6

Copyright © 2004 by Fredonia Books

Reprinted from the 1891 edition

Fredonia Books
Amsterdam, The Netherlands
http://www.fredoniabooks.com

All rights reserved, including the right to reproduce this book, or portions thereof, in any form.

In order to make original editions of historical works available to scholars at an economical price, this facsimile of the original edition of 1891 is reproduced from the best available copy and has been digitally enhanced to improve legibility, but the text remains unaltered to retain historical authenticity.

DEDICATED to the intelligent Druggists of America, in whose intelligence we firmly believe, trusting that in the near future this term will prove synonymous with every member of the profession.

PREFACE.

ONE NEED not expect that in the perusal of this book he will find formulas for all the fine flavors of the Orient, how best to mix the paste of almonds with sugar, or how orange blossoms are to be beaten with honey, nor how to place the sprig of mint to best flavor the cup of greenest tea, nor yet the proper mode of applying the water of roses to the already finely flavored tobacco of the East which the opulent Moor does most delight in. It lays claim only to Caucasian civilization.

In presenting this monograph, we hope to supply a legitimate want; namely, to furnish the druggists of America with a concentrated collection of facts on the composition and manufacture of Flavoring Extracts and Essences.

While not claiming this work to be of a scientific character, we would presume to a correct statement of facts; our formulas being put in such terms that there will be no doubt as to the exact meaning intended to be conveyed, and their intelligent comprehension made easy.

Its intent is to give the progressive druggist a proper and complete knowledge of the art of making Flavoring Extracts and Essences with their natural attendants, Syrups and Colorings according to advanced methods, and fully up to the best practice of the art of the present day.

Many of the formulas and facts herein given are the result of years of experience and labor, as well as, in part, a compilation derived from recent and reliable sources. We have used in its preparation the labors of such authors as are worthy of the highest confidence, and employed great care and diligence in the arrangement and selection of the materials gathered.

We would especially give credit to two names that will ever stand high in Amecican Pharmacy—Prof. Wm. Proctor, Jr., and Prof. Edward Parrish. To the latter we can refer with pride as having been our preceptor in Pharmacy. We are indebted also to Prof. P.

W. Bedford for many valuable hints, and last, but not least, to Prof. H. M. Whelpley for kindly advice; also to the United States Pharmacopœia and the Dispensatories, United States and National; Parrish's Practical Pharmacy, also Remington's Practice of Pharmacy; the Proceedings of the American Pharmaceutical Association, the National Formulary, the American Journal of Pharmacy, the Pharmaceutical Record, the Druggists' Circular, and the various druggists' journals of the day.

If our efforts shall not prove successful, it will be to us a source of regret; if the opposite shall be attained, we will be correspondingly happy.

J. H.

COLUMBUS, OHIO, January, 1891.

CONTENTS.

Introduction—Weights and Measures 17

PART FIRST.

Articles Used in the Manufacture of Flavoring Extracts.

Alcohol 22
Water 24
Essential Oils 25
Vegetable Aromatics 27

PART SECOND.

Flavoring Extracts.

Manufacture of Flavoring Extracts....... 31
Quality of Flavoring Extracts Sold 32
Vanilla Beans......................... 33
Quality of Vanilla Beans............... 34
Exhaustion of Vanilla Beans 36
To Make a Good Extract of Vanilla 38
Extract of Vanilla, *a, b, c, d, e,* and *f*. ..39–42

Tonka Beans 43
Extract of Vanilla with Tonka 44
Extract of Vanilla, "Standard" 46
Extract of Vanilla without Vanilla 47
Oil of Lemon 49
Lemon Extracts 50
Spirit of Lemon, U. S. 51
Tincture of Lemon, Br. 51
Extract of Lemon, *a* and *b*... .... 52, 53
Extract of Lemon, Improved............ 53
Extract of Lemon, "Standard"........... 54
Oil of Orange 55
Tincture of Sweet Orange Peel, U. S 56
Extract of Orange, *a* and *b* 57
Extract of Bitter Almond, *a*, *b*, and *c*... 58, 59
Extract of Rose, *a*, *b*, and *c* 59, 60
Extract of Nectarine 61
Extract of Cinnamon, *a* and *b* 61, 62
Extract of Nutmegs and Mace, *a* and *b*... 62
Extract of Cloves, *a* and *b* 63
Extract of Allspice 63
Extract of Ginger...................... 64
Extract of Black Pepper 64
Extract of Capsicum 64
Extract of Celery, *a* and *b* 65

FLAVORING EXTRACTS. 11

Extract of Pot or Soup Herbs, *a* and *b*...65, 66
Extract of Thyme 66
Extract of Sweet Basel 67
Extract of Sweet Marjoram 67
Extract of Summer Savory 67
Extract of Coriander 67
Extract of Teaberry or Wintergreen 68
Extract of Sarsaparilla, *a* and *b* 68, 69
Extract of Chocolate 69
Extract of Coffee 70

PART THIRD.

Flavoring Essences.

Flavoring Essences 73
Essence of Pineapple, *a* and *b* 75, 76
Essence of Strawberry, *a*, *b*, and *c* 76, 77
Essence of Raspberry, *a* and *b* 78, 79
Essence of Melon 79
Essence of Gooseberry.....................80
Essenge of Grape 80
Essence of Apple 81
Essence of Orange 81
Essence of Lemon 82
Essence of Pear 82

Essence of Black Cherry.................. 82
Essence of Cherry 83
Essence of Plum 83
Essence of Apricot 84
Essence of Banana 84
Essence of Peach 84
Essence of Currant 85
Acid Solutions 86

PART FOURTH.
Syrups.

Syrup 89
Syrup, U. S....................... 90
Syrup, thin 91
Syrup of Vanilla, *a*, *b*, and *c* 91, 92
Syrup of Lemon, U. S. 94
Syrup of Citric Acid, U. S............. 94
Syrup of Lemon, *a*, *b*, *c*, and *d* 95–97
Syrup of Orange, U. S. 97
Syrup of Orange, *a*, *b*, and *c* 98, 99
Syrup of Red Orange 99
Fruit Syrup Formulas 100
Fruit Syrups 101
Syrup of Raspberry, U. S............. 101

Fruit Syrup of Raspberry, *a* and *b* ...102, 103
Fruit Syrup of Strawberry, *a*............ 103
Fruit Syrup of Pineapple 103
Fruit Syrup of Strawberry, *b* 104
Fruit Syrup of Apricot................. 104
Fruit Syrup of Banana 105
Fruit Syrup of Peach 105
Fruit Syrup of Tamarind 105
Fruit Syrup of Plum................... 105
Fruit Syrup of Grape 106
Nectar Syrup, *a* and *b* 106
Sherbet Syrup, *a* and *b*........... 106, 107
Frambois Syrup 107
Maple Syrup.......................... 107
Cream Syrup, *a*, *b*, and *c*........... 107, 108
Nectar Cream Syrup 108
Orgeat Syrup 109
Syrup of Fruti Fru 109
Walnut or Hickorynut Cream Syrup.... 110
Chocolate Syrup, *a* and *b* 111
Coffee Syrup, *a*, *b*, *c*, and *d* 112, 113
Syrup of Egg Phosphate 113
Syrup of Acid Phosphates............. 114
Syrup of Ginger, U. S. 114
Syrup of Ginger, *a* and *b* 115

Syrup of Ginger Ale 115
Syrup of Capsicum 115
Syrup of Root Beer 116
Syrup of Sarsaparilla, *a* and *b*. 116
Syrup of Iron, Malt, and Phosphates ... 117

PART FIFTH.
Colorings.

Colorings 121
Fruit Colorings........................ 122
Solution of Carmine, N. F. 123
Solution of Cochineal, N. F............ 124
Tincture of Cochineal, Br. 124
Solution of Cochineal 125
Tincture of Cudbear Compound, N. F. 125
Tincture of Saffron, U. S. 126
Tincture of Safflower 126
Tincture of Turmeric 127
Solution of Caramel 127

PART SIXTH.
Appendix.

Appendix 131
Preservation of Lemons................ 132
Grating............................... 132

Restoring Essential Oils.................... 133
Adulteration of Oil of Bitter Almonds .. 133
Purification of the Oil of Bitter Almonds, 134
Simple Separating Funnel 134
Examination of Vanilla Beans........... 135
Estimation of Oil Present in Flavoring
 Extracts 136
Some Flavoring Extracts of the Market 137
Soluble Extracts from Volatile Oils 138
Soluble Extract of Lemon 139
Soluble Extract of Ginger, a and b 140
Solution of Acid Phosphates 141
Compound Phosphate Solution 143
Fruit Acid Solution 143
Soda Fountain "Mixtures".............. 144
    Calisaya Cordial.................... 144
    Tonic Hypophosphites 144
    Beef, Wine, and Iron 144
    Coca Tonic......................... 145
    Coca and Calisaya 145
Foam 145
Solution of Albumen..................... 146
Solution of Irish Moss 146
Solution of Gum Arabic 147
Tincture of Quillaia, N. F. 147

Compound Soda Foam 148
Ales, Beers, Wines, etc. 148
    Ginger Beer or Ale, *a* and *b* 149, 150
    Root Beer or Ale 150
    Spruce Beer, *a* and *b* 151
    Ginger Wine 152
    Mead 152
Conclusion 153
Index 155

Introduction.

As this work is supposed to be for the use of druggists, no description of weights or measures is thought necessary, nor scarcely a word of explanation, save to observe that when we say ounces or pounds we mean avoirdupois, the weight now official in the United States Pharmacopœia, as well as that used in commercial affairs.

When we speak of grains or drams we mean troy, the weight used universally in our business, for such amounts.

When referring to fluiddrams or fluidounces, we mean fluiddrams or fluidounces, wine measure; likewise, in speaking of pints or gallons.

In writing drops or minims, we will consider the terms interchangeable, as in very small quantities the variation between one and the other would be but trifling, and to drop would be much more convenient.

We have, from practice, found that a book of formulas is not complete unless interleaved. It often occurs that one wishes a smaller or a larger amount than the formula gives, or a slight variation in the proportions of some of the ingredients, for economic or other purposes, may be advisable. In either event, where a permanent memorandum is wished, it is much better to have a blank page attached than a loose leaf, liable to be lost.

PART FIRST.

Articles
Used in the Manufacture of
Flavoring Extracts.

"It must be remembered that both good ingredients and skillful manipulation are essential to success, with even the best of formulas."

PROF. H. M. WHELPLEY.

Flavoring Extracts.

ARTICLES USED IN MANUFACTURING.

In this, as in all practical operations, a proper knowledge of the articles to be used is of the first importance, and, we might add, is absolutely necessary to the intelligent preparation of Flavoring Extracts.

Much too little is known of the conditions necessary to produce them, by the average druggist. This then will be our first endeavor; to impress this fact is all important.

In order to properly apply this knowledge, a certain degree of thought and care is also necessary; indeed, this will hold good under any condition or circumstance, where mechanical operations are to be prosecuted, and much more so where substances so liable to change are used in the processes.

With these facts firmly fixed in the mind, the careful consideration of the articles used in the manufacture of Flavoring Extracts may be undertaken.

ALCOHOL.

Alcohol being the universal menstruum for the preparation of flavoring extracts, we will first consider the various kinds found on the market, and will call them:

First— Ordinary Alcohol.

Second— Middle Run or Deodorized (?) Alcohol.

Third— Cologne Spirit, True Deodorized or Atwood's Alcohol.

Ordinary Alcohol.

Ordinary alcohol is the alcohol in common use. It is the officinal alcohol of the pharmacopœia, of 94 per cent. strength, the same as is recognized by the United States government, the alcohol of commerce of to-day.

It is readily distinguished by its alcoholic(?) odor, which is due to the presence of fusel oil, and, although of the proper alcoholic strength, we consider it unfit for use in the manufacture of flavoring extracts.

Middle Run or Deodorized(?) Alcohol

This alcohol is that which is kept separate from the first and last which runs from the still in its manufacture, and contains much less fusel oil than either of those portions. It is far better suited for general use (except technical), than the ordinary alcohol. It is the article usually furnished by the wholesale trade when deodorized is ordered.

It can be used in the manufacture of flavoring extracts with moderate success, but is not pure enough for a first class product.

Cologne Spirit, True Deodorized or Atwood's Alcohol.*

This is the alcohol of alcohols. To get this is to get the best. Our advice is, be sure you get it. When mixed with oils or flavorings of any kind whatever, it is perfectly submissive. It does not speak up, by its presence, and say I am here, I, king alcohol; no, there is no fusel oil on which to base any such presumption. Your flavoring alone will speak; it alone has precedence.

NOTE—"*Perfumers' Alcohol* can now be had which, it is claimed, is very much superior to cologne spirit, as the purest alcohol obtainable was formerly called. It is termed perfumers' alcohol because it was found necessary to prepare a very high grade of alcohol for those who need a solvent for fine odors, on the score of economy, and to insure greater excellence of product."

* See page 1, Pubs. Dept.

WATER.

This article, from its name, may look quite thin and transparent, and, from a commercial standpoint, this is true; but when we look at it from an economic angle, we have quite a different conception of its value. The importance of water in the manufacture of flavoring extracts is next to alcohol, by replacing, in a degree, this valuable and expensive solvent, and at the same time serving the better to develop the flavoring principle. It, too, must be used with intelligence and care.

Water, for such purposes, is of two kinds:

First — Distilled Water.

Second — Ordinary Water.

Distilled Water.

This is by far the best kind to employ, but, as we all know, often inconvenient to get, especially in quantities, and still more difficult to keep, as liquids, more readily than solids, are prone to absorb impurities from the atmosphere; moreover, it is often, when purchased, no better than ordinarily pure water. For these reasons, it is not especially recommended.

Ordinary Water.

In speaking of ordinary water we may note the two varieties in common use, *soft* and *hard*. Either may be used, but the soft is to be preferred, for the lime and other mineral impurities held in solution in hard water are sure to precipitate when added to alcoholic liquids.

Boiling, and after standing a few hours, filtering, will much improve it. Soft water also is improved by the same treatment.

These casual remarks on this liquid and its use, in the preparation of flavoring extracts, will suffice, we trust, to give a proper conception of its value, and while apparently of no great importance, still the careful worker will appreciate and apply them.

ESSENTIAL OILS.

These are the most numerous and important constituents in the bases of flavorings, because of their concentrated form and cheapness of price, as well as the greater convenience in their application, as compared with the more crude conditions in which they originally exist. Their isolated and concentrated form is also their greatest danger. First, because of that enemy of oxidizable substances, the atmosphere; and, secondly, from that other and greater enemy (we hesitate to point squarely), we will say, the dishonest man. His generic name we know to be "mixer," and while we do not believe him to be amphibious, we are satisfied he can be found in most of our large cities. We have seen him at times ourselves, and the effect of his slight-of-hand performances we can never forget.

The proper selection and preservation of essential oils is a matter of no small importance, and until this feature of the work is well learned and conscientiously adhered to, there can be no certainty of securing the very end for which this book is written and which is, of all its features, the most important.

VEGETABLE AROMATICS.

Aromatic vegetable substances, from which flavoring extracts are made direct, are not numerous. The most important, perhaps, in every particular, being vanilla. This, as well as the liquid forms of the sources of flavorings, is liable to be deteriorated, not so much from natural as from causes directly under the control of, and often attributed to, man. A sorry fact, if true.

Other aromatic vegetable sources of flavorings are not to any great degree liable to sophistication.

In closing these remarks on the materials used in the manufacture of flavoring extracts, we have tried to outline a general plan of particulars which, we trust, will be an earnest of our efforts in the pages to follow. We shall attempt to lasso every fact of importance in the fundamental work, as well as in the detail of formula. If we fail to satisfy the most exacting, it will not be for want of earnest effort, and we trust our effort will not be in vain.

PART SECOND.

Flavoring Extracts

"The manufacture of flavoring extracts belongs properly to the art of pharmacy, but the business, through competition, has fallen into such hands that there is no longer any uniformity in the quality nor excellence in much that is made."

W. S. SNOW, PH. C.

Manufacture of Flavoring Extracts.*

An attempt at excuse for producing this monograph might, perhaps, be strengthened by reference to our text-books, especially our works on pharmacy. They tell of flavored syrups, but nothing is said of *flavoring extracts*. Hence, we see queries from druggists in pharmaceutical journals, asking for information regarding literature on this subject, and the reply that follows, "there is no work on flavoring extracts published, to our knowledge."

Thus we see that the manufacture of flavoring extracts, so far as our knowledge goes, is the only branch of industry which can be classed as an art, that has no written law by which it may be governed.

For every existing fact in nature there is said to be a cause; likewise, there may be good reasons for this existing fact.

---

*NOTE — The terms Concentrated Extracts, Concentrated Tinctures and Concentrated Essences, as referred to in price lists, often mean the same.

QUALITY OF FLAVORING EXTRACTS SOLD.

To make a good flavoring extract, requires great care, as well as a critical taste which will enable the operator to judge of the quality of the materials used.

It has been said that no careless man need attempt the business of wine making, for he will surely fail. We will say that in the manufacture of flavoring extracts no careless or ignorant man need attempt their preparation, for he will utterly fail.

We have on several occasions been asked by grocers our opinion of the quality of specimens of extracts offered for sale by manufacturers, and, as a rule, when the price was fair the extract was found to correspond with the price. Likewise, when a cheap article was offered, it was found invariably poor.

This rule, although holding good in the main, like all others, has its exceptions; especially is this so in cities, and more particularly in case of supplies furnished to confectioners, bakers and restaurants, when the article is sold in bulk. Men of much enterprise and little knowledge essay to enter a business and produce goods of which they know nothing. To secure business, they cut prices, and of necessity, they buy cheaply (the poorest is given them), common alcohol, often below the average, oils inferior, result, an extract unfit for use. We have seen such goods where the flavor was distinctly perceptible, but where the fusel oil was the more prominent of the two. While intending to make strong goods they, perhaps, put in the full amount of flavoring principle, but not understanding the conditions necessary to make a good extract, failed.

VANILLA BEANS.

The vanilla bean being the source of the most important and valuable flavoring which we have, will first be considered.

The bean-producing plant is a climbing parasite, the *Vanilla Planifolia*, of Andrews, according to the United States Pharmacopœia; but the bean of commerce is derived from various species of the genus *Vanilla*, according to our pharmaceutical writers. It is native to Mexico, the West Indies and South America.

and cultivated in the East Indies. That found on the market is of various kinds, as to name and quality. We have the Mexican, Bourbon, Imitation Mexican, Tahiti, Guatemala and Vanilon or Wild Vanilla, the latter differing most from the others in appearance, flavor and value.

Quality of Vanilla Beans.*

In quality we find quoted "ordinary," "fair," "good," "prime," "extra," "split," "cut," "broken lots of mixed lengths," and "powdered with fifty per cent. of sugar." Thus we have various grades, from which no one can fail to find his liking, either in price or quality. Prices ranging usually from two to twelve dollars a pound, often much higher.

The relative intrinsic value of vanilla beans, especially as to Mexican or Bourbon, appears to be subject to all the turnings of a theological question. "Make your extract of Mexican," "buy Bourbon," "try Tahiti," and the like advice can be met in druggist's journal or business circular, each claiming superiority or advantage for a certain kind, as in every

* See page ii, Pubs. Dept.

department of trade. But the old rule, "the quality regulates the price," will be a good one to remember in this, as in most other cases.

The question of length, as to value, which we never could fully comprehend, appears now to be ignored by some of the larger dealers. "We do not suppose that the mere length of the bean has any more to do with its flavor or flavoring qualities than the length or breadth of a man decides his mental or moral qualities. But as position and culture and education all play their part in the formation of character, so soil, climate and cultivation establish the quality of the fruit under consideration."

In purchasing vanilla beans it is almost a necessity to take them on trust, as to quality, although you are usually expected to pay for them in cash. This would appear an unreasonable condition of affairs and to be wholly objectionable, but it is only the first part of the conditions to which we take exceptions.

In buying vanilla beans try to deal, not merely with a reliable house, but with the most reliable vender of whom you have knowledge. To be candid, we need more light on the relative value of this fruit. All we are sure of in purchasing is the price and the length of the beans. They might be made of wood pulp, colored with caramel and flavored with synthetical vanilin, for all that.

Some years since we purchased a lot of vanilla, in which the coloring and aromatic principles were sadly deficient, so much so, that our reputation suffered before we were aware of the real facts; and the sorrowful part was, they were purchased from a good house, for a good bean, at a good price.

Exhaustion of Vanilla Beans.

The formula followed or the particular process to be adopted, does not appear to us of such vital importance as that whatever partic ular method be taken to secure the complete exhaustion of the bean, the work be well done, whether percolation, maceration or digestion, or a combination of the three.

The directions under each separate formula will, of course, apply to that formula, but the proper carrying out of the details can only rest with the operator.

Earnest, intelligent effort will always repay a hundred fold, in the manufacture of the extract of vanilla, as in every other process.

The proper menstruum, of course, will be the one that intelligent experiment has proven the most perfectly to exhaust and preserve the important principles of the substances operated upon.

The United States Pharmacopœia of 1880, under the name of *Tincture of Vanilla*, directs a menstruum composed of two parts alcohol and one part water (each by weight), of which fifty (50) parts are taken to ten (10) parts of the vanilla bean, and twenty (20) parts of sugar. The vanilla is cut in small pieces and macerated in half the mixture for twelve hours, the liquid drained off and set aside. The vanilla is then beaten into a uniform powder, with the sugar, in a mortar, packed in a percolator and the reserved liquid poured on; then the remainder of the liquid, and continued until one hundred parts of the "tincture" are obtained.

Prof. Wm. Proctor, Jr., has published the result of his efforts, as to the preparation of this particular extract, which we will give under the formula proposed by him.

We consider the vanilla bean as ranking among the hardest substances from which to extract its virtues, especially by percolation alone.

The following formulas will, however, speak for themselves.

To Make a Good Extract of Vanilla*

"The only requirements are cologne spirit, water, sugar, GOOD *beans and time, especially the last two. I have never yet been able to discover why brandy should be employed, except to increase the cost of the preparation; deodorized alcohol and water are quite as good, if not better. A mixture of cologne spirit, water and glycerin have been tried, but I have not found the addition of glycerin an improvement."* Dr. C. P. Nichols.

---

*This extract differs from most of the other important ones in its source, being made direct from the aromatic substance in its crude or natural condition; also in that it does not depend on a volatile oil for its virtue. This fact, together with experience, would seem to warrant the conclusion that time is a factor in the complete ripening or perfecting of this extract. You may demonstrate this by keeping an extract of vanilla for, say one year or over one summer, and then comparing with one recently made by exactly the same formula. This notion among manufacturing perfumers is accepted as a fact, as to the extract of musk.

Extract of Vanilla. * *a*

Vanilla (good quality) . . . . 1 ounce.
Sugar (coarse granulated) . . . . 2 ounces.
Simple Syrup . . . . 1 pint.

Diluted Alcohol, sufficient quantity.

Cut the vanilla, transversely, in small sections and triturate it with the sugar until reduced to a coarse powder; put this in a glass funnel prepared for percolation, and pour on diluted alcohol until a pint of tincture has passed; add this to the syrup and mix them.

It will be necessary to remark, with reference to this formula, that at the time of its publication (1866), the dilute alcohol in use among druggists and officinal in the United States Pharmacopœia, was of 39 per cent. strength; that is, equal parts of alcohol and water, by measure. The alcohol, too, was of 85 per cent. strength, consequently, the dilute alcohol of to-day would be a very much stronger spirit.

---

*Proctor.

NOTE—Flavoring Extract of Vanilla is sometimes erroneously called Fluid Extract.

Extract of Vanilla. * *b*

Vanilla Beans	1 pound.
Diluted Alcohol (Atwood)	2 gallons.
Alcohol (Atwood)	4 fl. ounces.

Cut the vanilla into very small fragments, and macerate in the diluted alcohol for two weeks; then place it in a displacement apparatus with an equal bulk of sand (washed); put the dilute alcohol through, and finally the strong.

The same remarks regarding the alcoholic strength may be applied to this formula as are made under the previous one.

Extract of Vanilla. † *c*
(QUICK METHOD)

Vanilla Bean (cut fine)	8 ounces.
Cologne Spirit	5 pints.
Water	3 pints.

Mix the liquids. Put one-third of the mixture in a suitable water bath apparatus with the cut beans. Cover closely, and heat to not over 140° F. for one hour, and remove the heat. Drain off the liquid, add another third of the

---
* Parrish.   † Bedford.

liquid, repeat the process, and again with the remaining portion of the liquid. Put the beans into a percolator, and having mixed one-half pint of the liquid in the proportions given, percolate to remove the last traces of the extract from the beans.

Filter the mixed liquids and pour the percolate on the filter to remove the adherent extract.

This will be found to be one of the most satisfactory of all processes, in the hands of a careful manipulator who knows how to avoid the risk of inflaming alcohol.

Extract of Vanilla. *d*

Vanilla .... 1 ounce.
Alcohol (95 per cent.)............. 3 fl. ounces.

Dilute Alcohol, sufficient to make 1 pint.

Cut the Vanilla into short pieces and bruise well with sand; then pack in a displacer; add first the strong alcohol, then the diluted alcohol, to make one pint. Let stand for twenty-four hours and filter. If desired, two ounces of syrup may be added to the gallon.

Extract of Vanilla. *e*

Vanilla Beans (Mexican) .... 4 ounces.
Sugar (granulated) .... 4 ounces.
Alcohol, a sufficient quantity.

Cut the Vanilla, transversely, into small pieces, and reduce it, with the Sugar, to as fine a condition as practical, by powdering in an iron mortar. Moisten the powder with 50 per cent. alcohol; pack in a percolator, in which allow the whole to macerate for twenty-four hours, and displace at the rate of 40 drops a minute, until four (4) pints of extract are obtained.

Extract of Vanilla. *f*

Vanilla Beans ... . 1 ounce.
Rock Candy .... 2 ounces.

Deodorized alcohol and water—a sufficient quantity of each.

Cut the Vanilla Beans in small pieces with a sharp knife, transfer to an iron mortar, and beat, with the Rock Candy, into a fine powder; place this in a bottle with nine (9) fluidounces of alcohol; allow to macerate, with occasional agitation, for twenty-four (24) hours, and add seven fluidounces of water; then treat in the same manner for two days, and filter.

TONKA BEANS.

(TONGA BEANS)

Tonka bean is the seed of *Diplerix Odorata*, of Wildinham, a large tree growing in Guiana. It is not described as being used for, or recommended as, a flavoring for culinary purposes, in the text-books, but only for flavoring snuff. We all well know, however, its extensive use as a flavoring in cookery.

Two varieties, as commonly found in the market, are noted, *Angostura* and *Para;* the former being held at a price much above the latter. Another kind, *Surianum*, is also known in commerce.

The Tonka bean has a strong, agreeable, rather heavy, aromatic odor, which, while not resembling the vanilla in flavor, is almost universally substituted for it in the manufacture of cheap forms of that extract, and accepted without question, from its long continued use, by a sane and confiding public.

The formulas for vanilla extract, which follow, will all contain it, to fill a want, and as a necessary condition of the trade, as we find it.

Extract of Vanilla with Tonka.

Vanilla Beans	4 ounces.
Tonka Beans	8 ounces.
Deodorized Alcohol (proof)	8 pints.
Simple Syrup	2 pints.

Cut and bruise the Vanilla Beans, afterward adding and bruising the Tonka Beans; macerate for fourteen days in one-half of the spirit, with occasional agitation; pour off the clear liquor and set aside; pour the remaining spirits on the magma, and heat by means of a water bath to about 170° Fahrenheit, in a loosely covered vessel; keep it at that temperature for two or three hours, and strain through flannel with slight pressure; mix the two portions of liquid and filter through felt; add the syrup.

If a genuine Extract of Vanilla is desired, take of vanilla beans six ounces, omit the Tonka, and proceed as above.

This process so exhausts the beans that percolation is unnecessary.

NOTE—The above process does not produce a perfectly clear extract. One-half dram of carbonate of magnesia to each ounce; rub well and filter: will produce a clear preparation.

Here we have an Extract of Vanilla with Tonka, and truly "there is no accounting for tastes," as the old lady with the bovine possession, remarked.

We once knew a lady of widely reputed good judgment and fine taste, who preferred an extract of vanilla made from a combination of vanilla and tonka to one made from the vanilla alone. The latter sample we know to have been pure and of good quality.

To quote from the author of the foregoing formula: "I even forgot that tastes differ, and that all do not smell from the same standpoint; that some who use the extract largely prefer one made from the vanilla bean, while others would select a preparation containing a certain proportion of the Tonka; that the dislike of some persons to vanilla, in any form, might lead them to pronounce the best extract inferior."

Hence, we say there can be no accounting for tastes. Lack of judgment or perception, it may be, often has more to do with like or dislike in such matters than anything else.

Extract of Vanilla — Standard.

Vanilla Beans .... 3 ounces.
Tonka Beans .... 6 ounces.
Sugar .... 12 ounces.
Alcohol (middle run) .... 1 quart.
Water .... 3 quarts.

Cut the Vanilla Beans, transversely, in small pieces and reduce to a fine powder, by placing in an iron mortar small quantities at a time, with two or three times the bulk of sugar; then reduce the Tonka beans to fine powder; mix well, pack firmly, without moistening, in a conical percolator; mix the liquids and percolate.

This formula represents a fair average of the respectable vanilla extracts of the market. We only state a fact, and will neither commend or condemn it.

Extract of Vanilla—without Vanilla.

Tonka Beans	10 ounces.
Prunes (freed from the seed)	1 pound.
Raisins	4 ounces.
Currants	3 ounces.
Orris Root (powdered)	4 ounces.
Balsam of Peru	3 ounces.
New Orleans Molasses	1 quart.

Alcohol and water, of each sufficient.

Bruise the Tonka Beans and digest for two or three hours in a quart of hot water. Cut the fruits small, add the powdered Orris, and cover with a mixture of alcohol, five pints, and water, one gallon. To this add the Tonka, both beans and the liquid; macerate for ten days; add the Balsam Peru and Molasses, and filter; lastly add enough diluted alcohol to make the extract measure two and one-fourth gallons, and color with solution of caramel, if desired.

Now, of the various formulas we have known, this one would appear to outdo them all. It is not true to name in the least particular, nor could we ask for excuse for placing it here, save as a curiosity, and to show what a formula may be.

"A great deal of conscientious care must be used in the selection of volatile oils, that they be of the best quality and recently distilled"

PROF. JOSEPH REMINGTON.

OIL OF LEMON.

This oil, from which the extract is produced, comes next in importance, as a flavoring, to vanilla, because so extensively used.

Oil of lemon is a volatile oil, separated by mechanical means from fresh lemon peel, which is the rind of *Citrus Limonum*, Risso; specific gravity, 0.850; soluble in two parts of alcohol. It may be preserved from the effects of oxidation by mixing it, while fresh, with five (5) per cent. of alcohol and separating the oil after it has become clear. Keep in a cool place.

As mankind is divided into different races so, commercially, oil of lemon is divided into different grades; these, however, unlike the former, only remain separate for a short period and do not, perhaps, continue so after they have left second hands. They are sometimes called "select," "extra," "prime," or "fair," with prices to suit the different kinds, higher or lower, as you ascend or descend the scale. In buying, buy the *best; good* may mean very little; the *best* is none too *good*.

Lemon Extracts.

This largely used extract, perhaps, from its extensive manufacture and sale, has suffered more abuse and misrepresentation than all the others combined, excepting vanilla, although its preparation, if the fundamental rules laid down in this work are observed, is not difficult. Its sale, in some parts of the country, is much in excess of the universally admired vanilla, while in others the reported demand is not greater.

We give, under this head, two formulas for the above named extract. The first as spirit, sometimes called Essence (?) of Lemon, of the United States Pharmacopœia; the second as Tincture of Lemon or Tincture of Fresh Lemon Peel, of the British Pharmacopœia. Both are good preparations. The U. S. formula contains the oil and rind, while the Br. has only the rind added to the spirit. Each is given in the language of its particular authority.

Spirit of Lemon — U. S.
(ESSENCE(?) OF LEMON)

Oil of Lemon .... 6 parts.
Lemon Peel (freshly grated) . ... 4 parts
  Alcohol, a sufficient quantity.

Dissolve the Oil of Lemon in ninety (90) parts of Alcohol, add the Lemon Peel, and macerate for twenty-four hours; then filter through paper, adding through the filter enough Alcohol to make the spirit weigh one hundred (100) parts.

Tincture of Lemon — Br.
(TINCTURE OF FRESH LEMON PEEL)

Fresh Lemon Peel (sliced thin) . 2½ ounces.
Proof Spirit* .... 1 pint, imp.

Macerate for seven (7) days in a closed vessel, with occasional agitation; strain, press and filter; then add sufficient Proof Spirit to make one (1) pint, imperial measure.

"Concentrated tinctures" of lemon and orange are now coming into use, which are sold as superior flavorings.

---

*Proof spirit (Br.), may be made by mixing alcohol (U. S.), 61 parts and water 42 parts, the mixture shrinking to 100 parts.

NOTE—A *Tincture of Fresh Lemon Peel* formula was published a few years since, in which it was directed to pare the fruit thinly and place it in a suitable vessel with deodorized alcohol, using four (4) ounces of peel to the pint. After standing for thirty (30) days, draw off and filter. This was called a saturated or stock tincture, to be used in making the Flavoring Extract.

Extract of Lemon. * *a*

Rind of Lemon (exterior) . . . . 2 ounces.
Alcohol (95 per cent.), deodorized, 2 pints.
Oil of Lemon (recent)............3 fl. ounces.

Expose the Lemon Rind to the air until perfectly dry, then bruise it in a wedgewood mortar and add it to the Alcohol, with agitation, until the color is extracted; then add the Oil, and, if it does not immediately dissolve and become clear, let it stand, with occasional agitation, for a day or two, and filter.

The color for this extract may be obtained from safflower, but, for many reasons, it is best to use the natural lemon color. The object of exposing the rind is to avoid weakening the alcohol, which should be as pure as possible.

When materials used are the best, and the extract is well corked in a full bottle, it improves by standing a few weeks before filtering.

---
* Proctor.

Extract of Lemon. † *b*

Oil of Lemon (fresh) .... 8 ounces.
Lemon Peel (fresh, grated) .... 4 ounces.
Alcohol (Atwood's, diluted, *q. s.*).. 1 gallon.

Mix Oil and Peel of Lemon with seven (7) pints of the alcohol, then add a mixture of water and alcohol, one (1) pint, in such proportions that the mixture will be only slightly clouded; let stand seven days and filter for use.

Extract of Lemon — Improved.

Oil of Lemon (select)............ 8 fl. ounces.
Oil of Lemongrass (fresh) . .... 1 fl. dram.
Lemon Peel (fresh grated) from. 1 dozen.
Alcohol (Atwood's) .... 7 pints.
Water (boiled)................. 1 pint.

Mix and macerate for seven days. If in a hurry for the product, percolate through the Lemon Peel and filter.

The addition of any other substance than the oil and rind of the lemon has not, so far as we know, been recommended. A circumstance that occurred some years since has led us, after the lapse of a decade, to the belief that an

---
†Parrish.

addition may be made with great improvement in the product. In this departure, we literally "go to grass" for our addition, but it is to Lemongrass. However, to use the words of Franklin, "for want of care" in using this flavor, one may easily overdo the thing. As the result of our experience, we may venture the statement that after its value has become generally known, no extract of lemon will be considered perfect without it. It stands related to lemon extract as musk to perfumes. It is a fastener, a developer, and while not made from the lemon, it is pre-eminently the thing.*

Extract of Lemon — Standard.

Oil of Lemon	3 fl. ounces.
Spirit of Lemon	6 fl. ounces.
Tincture of Turmeric	1 fl. ounce.
Alcohol (middle run)	6¼ pints.
Water (boiled)	20 fl. ounces.

Mix and filter, if necessary.

As was remarked under Standard Extract of Vanilla, this formula also may be taken as producing an average extract of the market.

---

*The lower the grade of spirit and lemon oil used, the greater its perceived virtue.

OIL OF ORANGE.

The oil of orange is the source of the flavoring of that name. It is a volatile oil, extracted by mechanical means from fresh orange peel, which is the rind of *Citrus Aurantium*, Risso; specific gravity, 0.860; it dissolves in two parts of alcohol.

Oil of orange is very prone to decomposition and acquires a disagreeable terebinthinate odor. It may be preserved by mixing, while fresh, with five (5) per cent. of alcohol and proceeding as in the case of oil of lemon; or better, perhaps, by shaking in one-fourth its volume of water, separating and mixing with five times its measure of alcohol. Keep in a cool place.

Several grades of this oil may be found in first and second hands, at prices to correspond. The same advice is given here as under oil of lemon, and even greater care should be employed in its selection than in the case of that oil. In buying, buy the best only; price should be a secondary consideration. It is the chameleon among volatile oils, and, although "change" is not printed on the label, it can be found as time goes on, by examining the contents of the bottle.

Tincture of Sweet Orange Peel-U. S.

Sweet Orange Peel (recently separated from the fresh fruit and deprived of the inner white layer)   20 parts.
Alcohol (sufficient to make).... ...100 parts.

Mix the Orange Peel, previously cut into small pieces, with eighty (80) parts of alcohol, and macerate for twenty-four hours; then pack it moderately in a conical percolator, and gradually pour alcohol upon it until one hundred (100) parts of tincture are obtained.

We think it would be much better to grate the orange peel, as a matter of neatness, as well as economy of time and perfection of process.

No authority, as obtained from the books, would warrant one in using any other than the ordinary alcohol, that in common use.

This delicate flavor would be a good one on which a progressive druggist could readily satisfy himself as to this point.

As observed under Tincture of Lemon, Br., flavorings called "concentrated tinctures," are coming into use.

Extract of Orange. * *a*

Rind of Orange (exterior) ... ... 2 ounces.
Alcohol (95 per cent., deodorized), 1 pint.
Oil of Orange .. .. 2 fl. ounces.

Proceed as in the recipe for Extract of Lemon. It is much more difficult to obtain oil of orange in a fit state for making this extract than that of lemon, and none should be used that is not perfectly free from the terebinthinate odor developed by exposure and age.

In purchasing the oil for this purpose, it should be put into small bottles, nearly full, closely sealed, and kept in a dark place.

Extract of Orange. *b*

Oil of Fresh Orange Peel....... 4 fl. ounces.
Peel of Fresh Orange (grated) . 4 ounces.
Alcohol (Atwood's, diluted, *q. s.*), 1½ gallons.

Mix the Oil and Peel of Orange with ten pints of the alcohol and proceed in the same manner as directed under formula for Extract of Lemon.

---
* Procter.

Extract of Bitter Almond. *a*

(EXTRACT OF PEACH)

Oil of Bitter Almond .... 2 fl. drams.
Alcohol (95 per cent., deodorized), 1 pint.
Tincture of Turmeric or Safflower, ½ fl. dram.

Mix and filter.

The directions accompanying this preparation should state that it is poisonous in quality.

It is not unusual in England to deprive the oil of bitter almonds, to be used in flavoring, of its hydrocyanic acid, before diluting it. As some may prefer to do this, to secure their preparation from the danger always incident to selling so potent a poison as the oil of bitter almonds for culinary purposes, even as a dilute solution, we offer a process for removing the poison.†

Extract of Almond. *b*

(EXTRACT OF PEACH)

Oil of Bitter Almond .... 1 fl. ounce.
Alcohol (Atwood's) .... 2 pints.
Water .... 4 pints.

Dissolve the Oil in the Alcohol and add the Water gradually, taking care not to make the solution milky.

†See Appendix.

Extract of Almond. *c*

(EXTRACT OF PEACH)

Oil of Bitter Almond .... 1 fl. dram.
Alcohol (deodorized)............ 10 fl. ounces.
Water (warm) .... 5 fl. ounces.

Mix the Oil with the Alcohol, and after allowing to stand twenty four (24) hours, add the Warm Water.

Extract of Rose. *a*

Oil of Rose .... ½ fl. dram.
Hundred-leaved Roses (recent)... 1 ounce.
Deodorized Alcohol .... 1 pint.

Bruise the Rose Leaves, extract by maceration in the Alcohol; follow by expression, so as to get a pint, in which dissolve the Oil, and filter.

In the absence of the recent rose leaves, dried red rose leaves may be used, or this ingredient may be omitted, adding a minute quantity of tincture of cochineal, to give a pale rose tint.

Extract of Rose. *b*

Oil of Rose (Kezanlick) .... ½ fl. dram.
Red Rose Leaves .... 2 ounces.
Alcohol (Atwood's) .... 2 pints.
Water .... 4 pints.

Dissolve the Oil in the Alcohol and gradually add the Water, then the Rose Leaves, and macerate for seven days and filter.

A great difference in the strength of these two formulas is observed.

Extract of Rose. *c*

Otto of Rose (Kezanlick) ....60 minims.
Alcohol (deodorized), and Water,
 of each. .... 8 fl. ounces.

Solution of carmine sufficient to color.
Mix and filter, if necessary.
The proportion of Otto may, of course, be varied. For a cheaper, but very good extract, half of otto of rose and half of otto of rose geranium may be used. Both should be of fine quality.

Extract of Nectarine.

Oil of Bitter Almonds . . . . . . 45 drops.
Oil of Lemon . . . . ¾ fl. ounce.
Oil of Orange . . . . ¾ fl. ounce.
Oil of Rose . . . . 8 drops.
Oil of Neroli . . . . 8 drops.
Tincture of Fresh Lemon Peel . . . 1 fl. ounce.
Tincture of Fresh Orange Peel . . . 1 fl. ounce.
Alcohol (deodorized) . . . . 40 fl. ounces.

Mix well and filter, if necessary. Color light red with tincture of cochineal.

Extract of Cinnamon. *a*

Oil of Cinnamon . . . . 2 fl. drams.
Ceylon Cinnamon (in powder) . . ½ ounce.
Alcohol (deodorized) . . . . 1 pint.
Water . . . . 1 pint.

Dissolve the Oil of Cinnamon in the Alcohol, and gradually add the Water, and then the Cinnamon, and agitate occasionally for several hours; lastly, filter the liquid through the dregs on a paper filter, so that it may be transparent. This preparation is much improved by using oil of Ceylon cinnamon, but when the oil of cassia is employed, the cinnamon powder partially corrects its flavor.

Extract of Cinnamon. *b*

Cinnamon Bark (true, in powder), 1 pound.
Alcohol (Atwood's, diluted)........1½ gallons.

Macerate for seven (7) days and filter.

It will be readily seen, by comparing this with the preceding formula, that though the color of the latter is much darker, the strength is much less, as is also the cost of manufacture.

Extract of Nutmeg or Mace. *a*

Oil of Nutmeg (of good quality)   2 fl. drams.
Mace (in coarse powder) .... 1 ounce.
Alcohol (deodorized) .... 2 pints.

Mix the Oil of Nutmeg and powdered Mace together, add them to the Alcohol, and, after several hours' maceration, filter the liquid through the dregs on a paper filter.

Extract of Nutmegs. *b*

Oil of Nutmegs............ ....   4 fl. ounces.
Nutmegs (grated) .... 15 in number.
Alcohol (Atwood's, diluted 1 to 2) 1 gallon.

Mix and macerate for seven (7) days and filter.

Extract of Cloves. *a*

Oil of Cloves .... 2 fl. drams.
Cloves (in coarse powder) .... 1 ounce.
Alcohol (deodorized) .... 2 pints.

Mix the Oil and powdered Cloves together, add them to the Alcohol, and, after several hours' maceration, filter the liquid through the dregs on a paper filter.

Extract of Cloves. *b*

Oil of Cloves .... 4 ounces.
Cloves (bruised) .... 1 ounce.
Alcohol (Atwood's, diluted, *q. s.*) ....1 gallon.

Mix and allow it to stand seven days and filter.

Extract of Allspice.

Oil of Allspice .... 2 fl. drams.
Allspice (in coarse powder) ....1 ounce.
Alcohol (deodorized) .... 2 pints.

Mix the Oil and powdered Allspice together, add them to the Alcohol, and, after several hours' maceration, filter the liquid through the dregs on a paper filter.

Extract of Ginger.

Jamaica Ginger (in fine powder) ... 4 ounces.
Simple Syrup .... ½ pint.

Alcohol (deodorized), a sufficient quantity.

Pack the Ginger, moistened with a little alcohol, in a funnel prepared for percolation, and pour on Alcohol until a pint and a half of tincture has passed; to this add the Syrup, and mix. If properly prepared, no precipitate occurs.

Extract of Black Pepper.

Black Pepper (in fine powder) ..... 4 ounces.
Alcohol (deodorized, sufficient to
    make) .... 2 pints.

Pack the Pepper, moistened with a little alcohol, in a funnel prepared for percolation, and pour on alcohol until two pints of tincture has passed.

Extract of Capsicum.
(EXTRACT OF CAYENNE)

Cayenne Pepper (in fine powder).... 4 ounces.
Alcohol (deodorized, sufficient to
    make) .... 2 pints.

Moisten, pack, and proceed as in formula for black pepper.

Extract of Celery. *a*

Celery Seed .... 2 ounces.

Alcohol, 95 per cent., deodorized, and water, each a sufficient quantity.

Bruise the Celery Seed finely, pack in a small percolator, and gradually pour on a pint of alcohol; then add water until first a pint of tincture and then a pint of infusion has passed; mix these, triturate with a dram of carbonate of magnesia, and filter through paper. As thus made, extract of celery has a light brown color, an agreeable odor and a well marked taste of celery.

Extract of Celery. *b*

Celery Seed (fresh crushed) .... 8 ounces.
Alcohol (Atwood's, diluted 1 to 2)...1 gallon.

Mix and macerate for seven days, and filter.

Extract of Pot or Soup Herbs. *a*

Thyme, Sweet Marjoram, Sweet Basil,
    Summer Savory (of each) ..... 1 ounce.
Celery Seed .... 1 dram.

Bruise all together until reduced to powder and percolate with sufficient diluted alcohol to make a pint of extract. The menstruum should be made with deodorized alcohol. Some prefer to add grated lemon peel, half an ounce, and either a little onion or garlic.

Extract of Soup Herbs. b

Summer Savory, Sweet Marjoram,
    Sweet Basil (of each)......... 2 tr. ounces.
Sage, Black Pepper (of each) .. ½ tr. ounce.
Thyme .... 1 tr. ounce.
Celery Seed .... 1½ drams.
    Alcohol and water, sufficient.

Reduce to a coarse powder, moisten with six fluidounces of a mixture of three and one-half (3½) pints of alcohol and one-half (½) pint of water; pack together in a percolator, and pour on the remainder of the menstruum. As soon as the liquid ceases to pass through, displace with diluted alcohol sufficient to make the product measure four (4) pints.

Extract of Thyme.

Thyme ... . 2 ounces.
Alcohol (Atwood's, diluted, sufficient
    to make) ... . 1 pint.

Bruise the Thyme well in an iron mortar, moisten it with one-half ($\frac{1}{2}$) fluidounce of the Diluted Alcohol; pack in a percolator, and pour Diluted Alcohol until a pint is obtained.

Extract of Sweet Basil.

Prepare in the same manner as the above.

Extract of Sweet Morjoram.

Prepare as in formula for thyme.

Extract of Summer Savory.

Prepare according to formula and directions given under extract of thyme.

Extract of Coriander.

Coriander (in powder)............ 4 ounces.
Oil of Coriander .... 1 fl. dram.
Alcohol (Atwood's, 95 per cent.)....1¼ pints.
Water .... ½ pint.

Mix the Alcohol and Water, then add the Coriander, previously mixed with the Oil, and macerate for twenty-four hours, with occasional agitation; finally, decant the liquid from the dregs, put these in a percolator, and pour on the decanted liquid; when this disappears, add sufficient diluted alcohol to make the percolate measure two (2) pints.

Extract of Teaberry or Wintergreen

Teaberries (ripe) .... 4 pints.
Oil of Teaberry .... 1 fl. ounce.
Alcohol (Atwood's, diluted, *q. s.*) 1 gallon.

Mix and macerate for several months, and filter.

Where dilute alcohol, *q. s.*, is directed in parenthesis, it refers to degree of dilution, as will be observed in formula for extract of lemon *b*.

Extract of Sarsaparilla. *a*

Oil of Anise, Oil of Sassafras (of
  each) .... 6¼ fl. drams.
Oil of Gaultheria .... 4¼ fl. drams.
Caramel Solution .... 1 fl. ounce.
Alcohol (deodorized, sufficient to
  make)...................... 2 pints.

Mix the Alcohol well with the oils and filter, if necessary, and add the Solution of Caramel.

Extract of Sarsaparilla. *b*

Oil of Wintergreen	6 fl. drams.
Oil of Sassafras	2 fl. drams.
Oil of Cassia	1½ fl. drams.
Oil of Cloves	1½ fl. drams.
Oil of Anise	1½ fl. drams.
Alcohol (sufficient to make)	8 fl. ounces.

Mix and filter, if necessary, and color to suit.

Extract of Chocolate.

Chocolate (powdered)*	4 ounces.
Syrup	5 fl. ounces.
Glycerin (pure)	6 fl. ounces.
Water (boiled)	3 fl. ounces.

Or a sufficient quantity.

Rub the Chocolate with the Glycerin and Syrup in a mortar until thoroughly mixed; transfer to a porcelain capsule, add the Water first to what adheres to the mortar, and transfer to the capsule; boil, with constant stirring, on a sand bath, for five minutes, and add water enough to make one pint. Flavor with extract of vanilla, if desired, bottle and cork well.

*See page iii, Pub's Dept.

Extract of Coffee.

Coffee (Java, roasted, No. 20 powder) .... 4 ounces.
Glycerin (pure) .... 4 fl. ounces.
Water and Boiling Water (of each sufficient to make) .... 1 pint.

Moisten the Coffee slightly with water and pack firmly in a tin percolator; pour on water gradually until four (4) fluidounces are obtained, and set aside; then place the coffee in a clean tin vessel with eight (8) fluidounces of water, and boil for five minutes. Again place the coffee in the percolator with the water, and when the liquid has passed or drained off, pack firmly and pour on boiling water until eight (8) fluidounces are obtained. When cold, mix with the first product and add the Glycerin; bottle and cork well.

The excellence of this extract of coffee, from the manner of its preparation, will be found by experience to be incomparably superior to that made by the formulas usually recommended, the reason being apparent in the first step of the process.

PART THIRD.

Flavoring Essences

Flavoring Essences.

FRUIT ESSENCES, ARTIFICIAL ESSENCES, ARTIFICIAL FLAVORS.

These Essences are intended to represent the flavoring principles of plants, and have come to be extensively used. They are sometimes called "extracts."

Fruit essences are made from combinations of ethers and alcohol, to which are sometimes added certain acids and natural "essences." Glycerin is present in almost all of them, and is added for the purpose of blending and harmonizing the various flavors. All the ingredients, alcohol as well, should be chemically pure.

These combinations have been made so perfectly to represent the natural fruit, that the food inspectors of Paris report that the only difference between the genuine and imitation is, that the latter appear to be the finer of the two.

Artificial flavorings, when properly prepared, are considered harmless.

As perfumes, they are not a success, producing headache and disagreeable symptoms. They somewhat resemble carbonic acid gas, which is the life of our carbonated beverages, and when taken into the stomach is healthful, but taken into the lungs to any great extent, is capable of producing fatal effects.

As to the manufacture of these goods by retail dealers, we note in a recent druggists' trade journal the following, which is a partial answer to a request to publish the formulas for the different artificial essences. After giving two or three formulas, they conclude: "Enough formulas have been given, we think, to show that the retail druggist will not care to make his own artificial fruit essences. The number of fruit ethers necessary to be carried is large and the stock expensive."

We will not discuss financial questions, as this work is not intended as a trade or market review; formulas and facts directly connected with them is our task; nevertheless, as we have promised all the facts, we will remark in this connection, that we have bought "concentrated extracts" of this kind where we would defy even Piesse himself to tell which was strawberry and which raspberry, except by reading the labels. They were exactly the same, save in color.

Since all wholesale druggists have become manufacturing chemists, there can be no guarantee for the quality of such goods.

Essence of Pineapple. *a*
(EXTRACT(?) OF PINEAPPLE)

Chloroform	1 part.
Aldehyd	1 part.
Butyric Ether	5 parts.
Amyl-Butyric Ether	10 parts.
Glycerin	3 parts.
Alcohol (deodorized)	100 parts.

Mix the Alcohol with all the ingredients, excepting the Glycerin; shake well, then add the Glycerin, and filter, if necessary.

It may be desirable to color the essence a light yellow, which may be done by adding a small quantity of tincture of turmeric.

These essences may be cheapened by replacing a portion of the alcohol with water.

Essence of Pineapple. *b*

Butyric Ether .... 1 fl. ounce.
Alcohol (deodorized) .... 1 pint.

Mix and color, if desired, and filter, if necessary. See under previous formula.

Essence of Strawberry. *a*

Nitrous Ether .... 1 part.
Acetic Ether .... 5 parts.
Formic Ether...................... 1 part.
Butyric Ether .... 5 parts.
Methyl-Salicylic Ether .... 1 part.
Amyl-Acetic Ether .... 3 parts.
Amyl-Butyric Ether .... 2 parts.
Glycerin .... 2 parts.
Alcohol (deodorized).............. 100 parts.

Mix and proceed as directed under essence of pineapple.

The appearance of the preparation may be improved by adding tincture or solution of cochineal or solution of carmine, a sufficient quantity.

Essence of Strawberry. *b*

Oil of Wintergreen	1 part.
Acetic Ether	5 parts.
Butyric Ether	5 parts.
Nitrous Ether	1 part.
Glycerin	2 parts.

Alcohol and water, each sufficient.

Mix sufficient of the above ingredients to make two (2) ounces, and add alcohol and water, equal parts, sufficient to make one (1) pint, and color to suit.

Essence of Strawberry. *c*

Butyric Ether	3 fl. drams.
Acetic Ether	3 fl. drams.
Nitrous Ether	$1\frac{1}{4}$ fl. drams.
Alcohol (deodorized)	1 pint.

Mix and add color to suit.

Another imitation may be made by adding to a weak solution of butyric ether, in alcohol, a very small proportion of oil of cloves.

It will be seen by comparing the three foregoing formulas that butyric and acetic ethers form the base, although the combination may be added to almost without limit.

Essence of Raspberry. *a*

Nitrous Ether	1 part.
Aldehyd	1 part.
Acetic Ether	5 parts.
Formic Ether	1 part.
Butyric Ether	1 part.
Benzoic Ether	1 part.
Œnanthylic Ether*	1 part.
Sebacic Ether	1 part.
Methyl-Salicylic Ether	1 part.
Amyl-Acetic Ether	1 part.
Amyl-Butyric Ether	1 part.
Tartaric Acid (saturated solution)	5 parts.
Succinic Acid	1 part.
Glycerin	4 parts.
Alcohol (deodorized)	100 parts.

Mix and proceed as directed under essence of pineapple *a*.

Color as under strawberry, but darker.

*Sometimes called Œnanthic Ether.

Essence of Raspberry. *b*

Butyric Ether	60 drops.
Acetic Ether	40 drops.
Nitrous Ether	10 drops.
Glycerin	20 drops.
Alcohol (deodorized)	2¼ fl. ounces.

Mix and proceed as directed under essence of pineapple. Color as under previous formula.

A simpler method for preparing the above essence consists in adding a small proportion of acetic ether to a strong tincture of orris root.

Essence of Melon.

Aldehyd	2 parts.
Formic Ether	1 part.
Butyric Ether	4 parts.
Valerianic Ether	5 parts.
Sebacic Ether	10 parts.
Glycerin	3 parts.
Alcohol (deodorized)	100 parts.

Mix and proceed as directed under essence of pineapple.

Color as directed under essence of strawberry, but the color should be of a shade between raspberry and strawberry.

Essence of Gooseberry.

Aldehyd	1 part.
Acetic Ether	5 parts.
Benzoic Ether	1 part.
Œnanthylic Ether	1 part.
Tartaric Acid (saturated solution)	1 part.
Benzoic Acid (saturated solution)	1 part.
Alcohol (deodorized)	100 parts.

Mix and filter, if necessary, and color with a small quantity of solution of caramel.

Essence of Grape.

Chloroform	2 parts.
Aldehyd	2 parts.
Formic Ether	2 parts.
Œnanthylic Ether	10 parts.
Methyl-Salicylic Ether	1 part.
Tartaric Acid (saturated solution)	5 parts.
Succinic Acid (saturated solution)	3 parts.
Glycerin	10 parts.
Alcohol	100 parts.

Mix and proceed as directed under essence of pineapple, and color with solution of caramel.

Essence of Apple.

Chloroform	1 part.
Nitrous Ether	1 part.
Aldehyd	2 parts.
Acetic Ether	1 part.
Amyl-Valerianic Ether	10 parts.
Oxalic Acid (saturated solution)	1 part.
Glycerin	4 parts.
Alcohol (deodorized)	100 parts.

Mix and proceed as under essence of pineapple, and color with tincture of turmeric.

Essence of Orange.

Chloroform	2 parts.
Aldehyd	2 parts.
Acetic Ether	5 parts.
Formic Ether	1 part.
Butyric Ether	1 part.
Benzoic Ether	1 part.
Methyl-Salicylic Ether	1 part.
Amyl-Acetic Ether	1 part.
Oil of Orange	10 parts.
Tartaric Acid (saturated solution)	1 part.
Glycerin	10 parts.
Alcohol (deodorized)	100 parts.

Mix and proceed as under essence of pineapple, and color with tincture of turmeric.

Essence of Lemon.

Chloroform	1 part.
Nitrous Ether	1 part.
Aldehyd	2 parts.
Acetic Ether	10 parts.
Oil of Lemon	10 parts.
Succinic Acid (saturated solution)	1 part.
Glycerin	5 parts.
Alcohol (deodorized)	100 parts.

Mix and proceed as directed under essence of pineapple, and color with tincture of turmeric.

Essence of Pear.

Acetic Ether	5 parts.
Amyl-Acetic Ether	2 parts.
Glycerin	2 parts.
Alcohol (deodorized)	100 parts.

Mix and proceed as directed under essence of pineapple, and color with tincture of turmeric.

Essence of Black Cherry.

Acetic Ether	10 parts.
Benzoic Ether	5 parts.
Oil of Persicot	2 parts.

Oxalic Acid (saturated solution)...... 1 part.
Benzoic Acid (saturated solution).... 2 parts.
Alcohol (deodorized) ....100 parts.

Mix well and filter, if necessary. Color with solution of caramel.

Essence of Cherry.

Acetic Ether .... 5 parts.
Benzoic Ether .... 5 parts.
Œnanthylic Ether .... 1 part.
Benzoic Acid (saturated solution)..... 1 part.
Glycerin .... 3 parts.
Alcohol (deodorized) ....100 parts.

Mix and proceed as directed under essence of pineapple. Color with tincture of cochineal.

Essence of Plum.

Aldehyd .... 5 parts.
Acetic Ether .... 5 parts.
Formic Ether .... 1 part.
Butyric Ether .... 2 parts.
Oil of Persicot .... 4 parts.
Glycerin .... 8 parts.
Alcohol (deodorized) ....100 parts.

Mix and proceed as directed under essence of pineapple. Color with compound tincture of cudbear.

Essence of Apricot.

Chloroform	1 part.
Formic Ether	10 parts.
Valerianic Ether	5 parts.
Œnanthylic Ether	1 part.
Amylic Alcohol	2 parts.
Amyl-Butyric Ether	1 part.
Tartaric Acid (saturated solution)	1 part.
Glycerin	4 parts.
Alcohol (deodorized)	100 parts.

Mix and proceed as directed under essence of pineapple. Color with tincture of saffron.

Essence of Banana.

Butyric Ether	10 parts.
Amyl-Acetic Ether	10 parts.
Glycerin	5 parts.
Alcohol (deodorized)	100 parts.

Mix and proceed as directed under essence of pineapple. Color with tincture of turmeric·

Essence of Peach.

Aldehyd	2 parts.
Acetic Ether	5 parts.
Formic Ether	5 parts.
Butyric Ether	5 parts.
Valerianic Ether	5 parts.

Oil of Persicot .... 5 parts.
Sebacic Acid (saturated solution) .. 1 part.
Amyl Alcohol .... 2 parts.
Glycerin .... 5 parts.
Alcohol (deodorized).............. 100 parts.

Mix and proceed as directed under essence of pineapple. Color with tincture of turmeric.

Essence of Currant.

Aldehyd .... 1 part.
Acetic Ether .... 5 parts.
Benzoic Ether .... 1 part.
Œnanthylic Ether .... 1 part.
Tartaric Acid (saturated solution) .... 5 parts.
Succinic Acid (saturated solution) .... 1 part.
Benzoic Acid (saturated solution) .... 1 part.
Alcohol (deodorized) .... 100 parts.

Mix well and filter, if necessary. Color with tincture of cochineal.

It will be observed that the relative strength, as given in the preceding formulas, varies considerably, and that the invariable "100 parts of alcohol" appears to be more of a fixed form, written by a scientific hand, than the practical work of a careful manufacturer.

It is apparent in these few remarks that license is given to figure for yourself, provided vou are able.

ACID SOLUTIONS.

The acids used in the above formulas are alcoholic solutions "saturated in the cold," which, by the way, is a comparative phrase; as cold as your store or shop would be liable to get, or at a temperature as low or lower than that to which your solution would be liable to be exposed; (the rule being, substances are less soluble in cold than in warm liquids). These alcoholic solutions would be liable to crystallize out if this precaution was not taken.

They are as follows:

Alcoholic Solutions of Acids (Saturated in the Cold).

> TARTARIC ACID,
> OXALIC ACID,
> SUCCINIC ACID,
> SEBACIC ACID.
> BENZOIC ACID.

Convenience only has caused this peculiar mode of preparing the above articles in solution, for the foregoing formulas, in which they are used.

PART FOURTH

Syrups.

"The use of plain syrup for diluting the stronger flavors is a necessity, and may be met by either making it direct from granulated sugar or the purchase of rock candy syrup. The latter is furnished, of an unexceptional quality and brightness and at a reasonable price, and labor is economized."

PROF. P. W. BEDFORD.

Syrup.

SIMPLE SYRUP.

Syrup is a concentrated solution of sugar in water or aqueous liquids.

The sugar to be used in making Syrup should be white, dry, hard, and in distinctly crystallized granules, permanent in the air.

Syrup may be made by solution, with heat; by agitation, without heat; or by percolation.

We have several varieties of Syrup, among which is rock candy syrup. Now, while we do not doubt the existence, in commerce, of such a syrup, we do fully believe that not one gallon in a hundred, sold as such, ever was in the condition of rock candy. Of course, we except goods sold by manufacturers of rock candy, who have, as a by-product or as drippings which have assumed a semi-amorphous condition, a genuine rock candy syrup.*

---

* See page v., Pub's Dept.

Syrup made with "C" sugar or a still poorer grade, by throwing into a large jar, with water, and stirring with a stick until dissolved, is not recommended for soda fountain or other use. We are assured, however, that such is the manner of manufacture as employed by some pharmacists.

The addition of antiseptics for the purpose of preserving thin syrups is not desirable, and all impaired or sour syrups should be disposed of by way of the drain.

Syrup—U. S.

(SIMPLE SYRUP)

Sugar .... 80 ounces = 5 pounds.
Water (distilled) .... 40 fl. ounces = 2½ pints.

Dissolve the Sugar, with the aid of heat, in the distilled Water; raise the temperature to the boiling point, and strain the solution while hot; then incorporate with the solution enough distilled water, added through the strainer, to make the syrup measure five (5) pints and ten (10) fluid ounces.

Syrup—Thin.

(THIN SIMPLE SYRUP)

Sugar .... 7 pounds.
Water (boiled and filtered) .... ½ gallon.

Mix and dissolve by heat.

Syrup of the strength produced by the above formula is recommended for use at the soda fountain, as syrup of the full officinal strength is too thick to mix readily with the soda water, and is inclined to adhere to the glass. Moreover, the bulk given by a diluted syrup does please the eye for quantity. This, however, must not be carried to the extreme, as is sometimes done.

Syrup of Vanilla. *a*

Extract of Vanilla .... 2 fl. ounces.
Syrup (sufficient to make) ...... 2 pints.

Mix well.

This syrup of vanilla would appear to be much stronger than that in common use. It is also without the coloring commonly added.

Syrup of Vanilla. *b*

Extract of Vanilla .... 2 fl. ounces.
Syrup (thin, enough to make)... 4 pints.

Mix well.

The addition of solution of caramel would add to the appearance of the syrup.

Syrup of Vanilla. *c*

Extract of Vanilla .... ½ fl. ounce.
Solution of Caramel .... ¼ fl. ounce.
Solution of Albumen .... 4 fl. ounces.
Syrup (thin, enough to make).... 2 pints.

Mix well.

A good syrup of vanilla is made by the above formula. The remarks as to strength, made under formula for vanilla *a*, will not apply to this.

The list of formulas for syrup of vanilla would not appear to be complete without one to which Tonka extract is added; this, however, can, and perhaps will be, done, without any specific directions from us.

"The formula for Syrup of Lemon, as directed in the Pharmacopœia of 1870, is far preferable to that of 1880. It keeps perfectly, and is a handsomer preparation. It is less acid, and we think, on that account, has a finer flavor."

H. O. RYERSON.

Syrup of Lemon — U. S.

Lemon Juice (recently expressed
  and strained) .... 17 fl. ounces.
Fresh Lemon Peel .... 1 ounce.
Sugar (in coarse powder) .... 28 ounces.
Water (a sufficient quantity to
  make about) .... 2 pints.

Heat the Lemon Juice to the boiling point, then add the Lemon Peel, and let the whole stand, closely covered, until cold; filter, add enough water, through the filter, to make the filtrate measure seventeen (17) fluid ounces; dissolve the Sugar in the filtered liquid, by agitation, without heat, and strain.

Syrup of Citric Acid — U.S.

Citric Acid .... 150 grains.
Water .... 3 fl. drams.
Spirit of Lemon .... 100 minums.
Syrup .... 2 pints.

Mix the Spirit of Lemon with the Syrup contained in a bottle; then add gradually the Citric Acid, dissolved in the Water, shaking the bottle after each addition until the whole is thoroughly mixed.

Syrup of Lemon. *a*

Solution of Citric Acid (1 to 10).. 3 fl. ounces.
Spirit of Lemon ....1½ fl. ounces.
Syrup .... 8 pints.

Tincture of turmeric sufficient to color.

Mix well.

Syrup of Lemon. *b*

Oil of Lemon .... 20 drops.
Citric Acid .... 1 ounce.
Tartaric Acid .... 2 drams.
Syrup........................ 1 gallon.

Sugar and water, of each sufficient.

Rub the Oil of Lemon with a little sugar and afterwards with a portion of the syrup, and having dissolved the Acids in a gill of water, mix the whole thoroughly together.

"This syrup is now almost universally made from citric or tartaric acid and oil of lemon, instead of lemon juice, and is superior to that sometimes made from inferior lemons.

"Citric acid is preferable to tartaric acid for preparing the syrup; when made from the former acid it has a more agreeable taste, which it retains longer unimpaired.

"The syrup made with either acid, when kept long, is liable to throw down a white granular deposit of grape sugar. A 'turpentine taste' is very common in the lemon syrup which is manufactured and sold wholesale, and may be frequently due to the employment of old or impure oil of lemon."

Syrup of Lemon. *c*

Lemon Peel (fresh); Alcohol (deodorized); of each, equal parts by weight.

Mix and macerate for twenty-four (24) hours in a covered vessel, after which the alcohol is drawn off by distillation.

This spirit of lemon is used by adding—

Spirit of Lemon .... 30 parts.
Syrup .... 750 parts.
Orangeflower Water .... 30 parts.
Citric Acid .... 15 parts.

Dissolve the Citric Acid in the Orangeflower Water and mix all well together.

Such a lemon syrup is said to be far superior, both in flavor and durability, to that made either from the freshly expressed juice or from citric acid and oil of lemon.

Syrup of Lemon. *d*

Lemons (select) .... 1 dozen.
Water (hot) .... 4 pints.
Sugar .... 6 pounds.

Cut the Lemons and bruise in a wedgewood mortar; add the Hot Water, let stand at a very gentle heat for twenty (20) minutes; add the Sugar, dissolve, express, and make up to one (1) gallon by addition of *thin* syrup. This syrup must not be used with cream.

By the same process, and in the same proportions, an excellent orange fruit syrup may be made by the above formula for lemon, substituting orange fruit.

Syrup of Orange—U. S.

Sweet Orange Peel (fresh, deprived of the inner white layer and cut in small pieces)........ 2¼ ounces.
Alcohol (deodorized) .... 3 fl. ounces.
Precipitated Phosphate of Calcium ¼ ounce.
Sugar. .... 28 ounces.
Water (sufficient to make)........ 2 pints.

Macerate the Orange Peel with the Alcohol for seven (7) days, then express the liquid.

Rub this with the Precipitated Phosphate of Calcium and thirteen (13) fluid ounces of water, gradually added; filter the mixture, and pass through the filter enough water to make seventeen (17) fluid ounces. Lastly, add the Sugar, dissolve it by agitation, without heat, and strain.

Syrup of Orange. *a*

Oil of Orange (fresh) .... 10 drops.
Citric Acid....................... ¼ ounce.
Syrup .... 4 pints.

Rub the Citric Acid with the Oil, then with the Syrup. Mix well and color with tincture of turmeric, if desired.

Syrup of Orange. *b*

Syrup of Orange (U. S.)............ 1 pint.
Citric Acid. ... ...45 grains.

Dissolve the Citric Acid in the Syrup of orange and color with tincture of turmeric, if desired.

Syrup of Orange. *c*

Take of oranges, the fresh fruit, a convenient number, grate off the yellow outside peel, cut the oranges and express the juice; to each quart of which add—

Water . ... 1 pint.
Sugar .... 6 pounds.

Mix the Sugar and the grated orange peel, add the mixed Water and orange juice, and apply a gentle heat until the Sugar is dissolved, then strain.

One dozen oranges will make from one and a half to two gallons of syrup.

Syrup of Red Orange.

This syrup may be made from the red variety of orange by the same process as given in the above formula. We think, however, it is frequently made by adding some coloring matter to the ordinary orange syrup. Even the red oranges themselves are said to be often colored by pricking and injecting a solution of red aniline.

Fruit Syrup Formulas.*

"From reliable fruit juices fruit syrups may be made for immediate use by mixing the contents of a bottle with three or four times its bulk of dense simple syrup or rock candy syrup. Beyond this point of dilution the dealer may go to such an extent as he chooses, but the smaller cost is offset by the disappointment of the consumer of the beverage, and we urge that a full, good flavor should not be sacrificed.

"Concentrated syrups from fruit juices will best suit those who do a small business; for this purpose, take the contents of a bottle of the juice, and weighing it, add one and three-fourths the weight of sugar and cause it to dissolve, using but little heat. When used for the soda fountain, add the same bulk of simple syrup or rock candy syrup, and to the mixture one-fourth the bulk of boiled and filtered water."

The fruit syrup formulas which follow are, for the most part, the result of long experience, and may be relied on as correct in every particular. The third name referred to in the preface being authority for most of them.

---
* See page iv., Pub's Dept.

Fruit Syrups.

To make one (1) gallon of raspberry, strawberry or blackberry syrup—

Take of the Fresh Fruit .... 4 quarts.
Sugar .... 6 pounds.
Water, a sufficient quantity.

Express the juice and strain, then add Water until it measures four (4) pints; dissolve the Sugar in this by the aid of heat, raise it to the boiling point, and strain. If it is to be kept until the following season, it should be poured, while hot, into dry bottles, filled to the neck, and securely corked and sealed.

These syrups contain a small quantity of alcohol, and keep well in sealed bottles, but exposed to the air, they soon undergo acetous fermentation.

Syrup of Raspberry — U.S.

The U. S. formula for syrup of raspberry produces a similar preparation, but contains more alcohol than the preceding one.

The additional precaution is added here, however, to avoid the use of tinned vessels, and keep in a cool and dark place.

Fruit Syrup of Raspberry. *a*

This may be made by preserving the fruit as follows: Three (3) quarts of raspberries are pulped (mashed) with an equal weight of sugar, heated by water bath in fruit jars, and sealed. When wanted for use, open this quantity, mix thoroughly with enough thin syrup to make one (1) gallon, and strain.

When fruit juice or fresh fruit cannot be had, and a fine quality of canned fruit is obtainable, the contents of a can may be pulped, heated gently and strained; if necessary, it may have more sugar added to make a denser syrup to keep it better for stock, and when wanted for use, dilute with water or thin syrup before placing in the fountain.

The two foregoing formulas are given for raspberry, but as will be evident on thought, any fruit capable of being treated in the manner given in the above formulas can be prepared by the same processes. We note these facts so that they may not be overlooked, and give samples only to economize space.

Fruit Syrup of Raspberry. *b*

Raspberry Juice .... 32 fl. ounces.
Sugar .... 128 ounces.
Water .... 32 fl. ounces.

Mix the Raspberry Juice and Water, and dissolve the Sugar, by percolating, with the mixture.

Fruit Syrup of Strawberry. *a*

Strawberry Juice .... 32 fl. ounces.
Sugar .... 128 ounces.
Water .... 32 fl. ounces.

Mix the Strawberry Juice and Water, and dissolve the Sugar, by percolation, with the mixture.

Fruit Syrup of Pineapple.

Pineapple Juice .... 32 fl. ounces.
Sugar .... 128 ounces.
Water .... 32 fl. ounces.

Mix the Pineapple Juice and Water, and dissolve the Sugar, by percolating, with the mixture.

The above three are samples only of the many that may be prepared in a similar way.

Fruit Syrup of Strawberry. *b*

Strawberry Juice.................... 1 pint.
Sugar .... 24 ounces.

Syrup, sufficient quantity. Solution of citric acid, the same.

The strawberry juice, in the above formula, is prepared by taking a sufficient quantity of the fruit, properly picking and cleaning, then covering slightly with sugar. Allow to stand for twelve (12) hours, express; add to one (1) pint the Sugar; dissolve with heat; strain and bottle while hot, and keep in a cool place. When wanted for use, add an equal bulk of plain syrup and a small quantity of solution of citric acid.

The same method may be used for all other fruits from which the concentrated syrups are prepared.

Fruit Syrup of Apricot.

Take of apricot paste* and water equal parts; heat gently, then add as much more water; continue the heat for a few moments, strain to remove the coarser portions of the pulp, and add to the liquid one and one-half its weight of sugar.

*Imported Apricot Paste is to be found on the market.

Fruit Syrup of Banana.

To each pound of banana pulp add gradually the same weight of hot water, heat gently; strain and add sugar three (3) pounds.

Fruit Syrup of Peach.

The pulp of ripe peaches is thoroughly mixed* with its own weight of water, gradually added; then pass through a moderately coarse strainer; to each quart add three (3) pounds of sugar, and dissolve.

This syrup may be closely approached in flavor by adding to apricot syrup one (1) quart, strawberry juice four (4) to six (6) ounces.

Fruit Syprup of Tamarind.

From select pulp of tamarinds, by the same formula as for peach.

Fruit Syrup of Plum.

This is frequently made by treating selected prunes with hot water, to extract their flavor and a portion of the pulp, and made as banana. But it is better made direct from the ripe fruit and acceptably from canned fruit.

*Best done by using a Keystone beater.

Fruit Syrup of Grape.

This syrup is made from the unfermented grape juice by adding syrup. It is agreeable and there can be no objection to its use, as is sometimes urged when brandy is used.

Nectar Syrup. *a*

Vanilla Syrup .... 40 fl. ounces.
Pineapple Syrup .... 8 fl. ounces.
Strawberry Syrup .... 16 fl. ounces.
  Mix well.

Nectar Syrup. *b*

Pineapple Syrup.................. 1 part.
Lemon Syrup..................... 1 part.
Vanilla Syrup .... 3 parts.
  Mix well.

Sherbert Syrup. *a*

Vanilla Syrup................. 48 fl. ounces.
Pineapple Syrup.............. 16 fl. ounces.
Lemon Syrup................. 16 fl. ounces.
  Mix well.

Now, although this formula and the previous one are from equally good authority, they will not bear close comparison as to difference.

Sherbert Syrup. *b*

Orange Syrup .... 1 part.
Pineapple Syrup .... 1 part.
Vanilla Syrup .... 1 part.

Frambois Syrup.

Raspberry Syrup .... 1 pint.
Currant Syrup................ 2 pints.
   Mix well.

The various fruit syrups, mixed, give rise to many other names.

Maple Syrup.

Maple Sugar (pure) .... 3 pounds.
Water .... 2 pints.
   Mix and dissolve by a gentle heat.

If the syrup can be obtained pure (which it is hard to do), it may be mixed with an equal bulk of simple or rock candy syrup.

Cream Syrup. *a*

Cream .... 1 pint.
Milk......................... 1 pint.
Sugar........................ 1 pound.
   Mix, dissolve without heat.

If this mixture is bottled at once and kept upon ice, it will keep well for from four to eight days.

Cream Syrup. *b*

Condensed Milk (without sugar)....1 pint.
Water (previously boiled and cooled) 1 pint.
Sugar ....1½ pounds.

Mix and dissolve without heat.

Cream Syrup. *c*

Condensed Milk (with sugar)...1 can or ½ pint.
Water (previously boiled and
  cooled................. ... ½ pint.
Syrup (thin) .... 1 pint.

Mix and dissolve without heat.

Nectar Cream Syrup.

Cream Syrup .... 6 pints.
Vanilla Syrup .... 3 pints.
Pineapple Syrup .... 1 pint.
Lemon Syrup .... 1 pint.

Mix well and color with tincture of cochineal, a sufficient quantity.

Orgeat Syrup.

Cream Syrup...................... 1 pint.
Vanilla Syrup..................... 1 pint.
Oil of Bitter Almond (or extract, 2 fl.
 drams)........................... 4 drops.

Mix well together and observe not to make more than sufficient for one day's sales, unless precautions given under cream syrup be observed.

Syrup of Fruti Fru.

Extract of Orange............... 4 fl. drams.
Extract of Lemon............... 6 fl. drams.
Extract of Vanilla............... 4 fl. drams.
Solution of Citric Acid .... 3 fl. ounces.
Syrup (thin).................... 1 gallon.

Solution of caramel and solution of cochineal, of each a sufficient quantity to produce a healthy color.

Mix the flavoring extracts with the syrup and afterward add the solutions and mix well.

Walnut or Hickorynut Cream Syrup

Take one (1) pound of hickory-nut or walnut kernels and remove the skin by blanching, which, if left on, would give an unpleasant, bitter taste; then powder in a wedgewood or porcelain mortar, adding a few drops of lemon juice to prevent the separation of the oil in kernels; also water, gradually added, to make a thick emulsion. As fast as the kernels are reduced, put them in a linen cloth, which should be gathered around them, so that they may be squeezed through the cloth. Whatever is left in the cloth is to be returned to the mortar and pulverized further; the lemon juice and water being added as needed. All should eventually pass through the strainer.

The result of this process, about two (2) pints, is to be added to two (2) quarts of cream syrup.

This formula may be varied, and perhaps improved, by a slight addition of extract of lemon or vanilla, or any other flavor to suit the taste; likewise a little coloring to suit the fancy. It will well repay the labor of preparing it.

Chocolate Syrup. *a*

Chocolate (powdered) .... 8 ounces.
Sugar .... 64 ounces.
Water .... 32 fl. ounces.

Mix the Chocolate with the Water and stir thoroughly over a slow fire, at boiling point, for a few minutes; strain; add the Sugar and dissolve.

Chocolate Syrup. *b*

Chocolate (powdered) .... 1 pound.
Water .... 4 pints.
Sugar .... 4 pounds.
Extract of Vanilla............. 1 fl. ounce.
Extract of Cinnamon .... ¼ fl. ounce.

Mix the Chocolate and Water well together in a mortar; transfer to a porcelain-lined kettle; add the Sugar; bring to the boiling point, with constant stirring; remove from the source of heat; continue the stirring for some minutes; when cold, add the Extract of Vanilla and Extract of Cinnamon and enough Syrup to make one (1) gallon.

*Much depends on the proper selection of the chocolate used. See p. iii, publ.sher's department.

Syrup of Coffee. *a*

Java Coffee (ground very fine)   2 pounds.
Sugar .... 4 pounds.
Alcohol (deodorized) .... 2 pints.
Water .... 6 pints.

Moisten the Coffee and pack in a suitable percolater; add the remaining liquid to thoroughly exhaust it. At a very gentle heat evaporate the Alcohol and add the Sugar. Make to the measure of one (1) gallon by adding thin, simple syrup,

---

Syrup of Coffee. *b*

Mocha Coffee .... 4 ounces.
Java Coffee .... 4 ounces.
Sugar........................... 7 pounds.
Water, Boiling, a sufficient quantity.

To the mixed coffee, first slightly moistened and packed in a tin percolator, add the Boiling Water until one-half ($\frac{1}{2}$) gallon of the product is obtained; in this dissolve the Sugar and strain, if necessary.

Syrup of Coffee. c

Coffee (roasted and ground) ..... 8 ounces.
Boiling Water .... 8 fl. ounces.
Sugar .... 120 ounces.

Make an infusion, filter, add the sugar, dissolve and strain, if desired.

Syrup of Coffee. d

Extract of Coffee............... 4 fl. ounces.
Syrup ....12 fl. ounces.
Mix well.

Syrup of Egg Phosphate.

Lemon Syrup .... 2 pints.
Orange Syrup.............. 2 pints.
Eggs .... 2⅔ dozens.
Phosphoric Acid (U. S.) ... 1 to 2 fl. ounces.

Thoroughly incorporate these ingredients with a Keystone beater or other suitable means.

Syrup of Egg Phosphate.
(For Single Glass.)

Lemon or Orange Syrup. ... 1 to 1½ fl. ounces.
Compound Phosphate Solution .... 1 fl. dram.

Shaven Ice............ ..... 2 ounces.
Eggs.. .... ₁ dozen.
Water (iced).. ... . 2 fl. ounces.

Mix well by shaking vigorously; strain into a tumbler and fill up with carbonated water.

Syrup of Acid Phosphates.*

Solution of Acid Phosphates.. ... 8 fl. ounces.
Syrup...... .... 7½ pints.

Mix and flavor as desired.

Syrup of Ginger—U. S.

Fluid Extract of Ginger . ... 1 fl. ounce.
Sugar (in coarse powder)........30 ounces.
Water (sufficient to make about).. 2 pints.

Rub the Fluid Extract with twelve (12) ounces of sugar, and expose the mixture to a heat of not exceeding 140° F., until the alcohol is evaporated; then mix the residue thoroughly, by agitation, with fifteen (15) fluidounces of water, and filter the liquid, adding, through the filter, enough water to make the whole measure twenty-two (22) fluidounces; finally, add the remainder of the sugar, dissolve it by agitation, without heat, and strain.

---

*This formula will answer for either simple or compound.

Syrup of Ginger. *a*

Soluble Extract of Ginger .... 2 fl. ounces.
Syrup (sufficient to make)........ 4 pints.
 Mix well.
 This formula affords a delicate and pleasant flavor. If a syrup of more pungency is desired a small quantity of extract of capsicum may be added.

Syrup of Ginger. *b*

Tincture of Ginger (U. S.) or Extract of Ginger (flavoring).... 4 fl. ounces.
Syrup......................... 8 pints.
 Mix well.

Syrup of Ginger Ale.

Ginger Syrup .... 2 pints.
Extract of Lemon.............. ½ fl. ounce.
Solution of Citric Acid .... ¼ fl. ounce.
 Mix well.

Syrup of Capsicum.

Extract of Capsicum .... 1 fl. ounce.
Syrup .... 2 pints.
 Tincture of cochineal, sufficient.
 Mix well.

This syrup is recommended as a good stimulant, which may be used to advantage in soda water, more especially in hot and debilitating weather, when the relaxed condition of the system and especially of the digestive organs would seem to contra-indicate the use of cold drinks.

Syrup of Root Beer.

Root Beer Extract .... 1 to 2 fl. ounces.
Thin Simple Syrup .... 4 pints.

Mix well and color with solution of caramel to suit.

Syrup of Sarsaparilla. *a*

Compound Syrup of Sarsaparilla
 (U. S.) .... 4 fl. ounces.
Thin Simple Syrup .... 4 pints.
Solution of Caramel .... 2 fl. ounces.
Extract of Wintergreen .... 1 fl. dram.
Extract of Sassafras .... 1 fl. dram.
 Mix well.

Syrup of Sarsaparilla. *b*

Fluid Extract of Sarsaparilla
 (U. S.) .... 2 fl. ounces.
Oil of Sassafras ....12 minims.
Oil of Anise....................12 minims.

Oil of Gaulthera .... 9 minims.
Solution of Caramel .... 4 fl. ounces.
Thin Simple Syrup....... .. ... 8 pints.

Mix the Oils in a little alcohol, add them to the Syrup, shake thoroughly, and add the Fluid Extract of Sarsaparilla.

Syrup of Iron Malt and Phosphate.

Solution of Pyrophosphate of Iron
  (1 to 8) .... 2 fl. drams.
Extract of Malt .... 1 fl. ounce.
Solution of Acid Phosphates ... 1 fl. ounce.
Solution of Albumen........... 2 fl. ounces.
Solution of Caramel .... 2 fl. drams.
Extract of Vanilla............. 1 fl. dram.
Extract of Bitter Almonds ...... $\frac{1}{2}$ fl. dram.
Syrup (sufficient to make) ..... 20 fl. ounces.

Mix well.

This amount is sufficient to fill a bottle similar to those commonly used for such goods.

The formula above given is not placed in this work as of superior value or excellence in itself, but rather as an example or type of an extensively advertised and widely used class of goods. It will, however, we believe, be appreciated as a very satisfactory and saleable article.

PART FIFTH

Colorings.

Colorings.*

The proper employment of colorings in the various formulas given in this work is of considerable importance, more especially the kinds fit or best fitted to be used.

Our observation, extending through some years, has led us to the belief that very many of the flavoring extracts of the market are colored with aniline. These substances (the various salts of the base aniline), without doubt contain a poison, which in many instances has been found, on analysis, to be arsenic, and which has been said by expert chemists to be much more active in combination with aniline

---

*Juices, Syrups or Extracts, colored with Fuschin, color a woolen or silk thread permanent; natural fruit colors will wash out.

Dilute Mineral Acids redden natural fruit colors, turning yellow those containing aniline coloring.

Half a volume of Nitric Acid instantly turns an artificial color yellow.

Carbonate of Potassium reddens artificial colorings, but has no effect on natural.

Subacetate of Lead precipitates red with Fuschin, and green with natural fruit juices.

than when in a separate condition. The facts, as stated, appear to have led to such a conclusion. The inducement for such a substitution can only be accounted for on the grounds of cheapness. The fact admitted, the injurious effects proven, the cause economy only, then we need not ask for excuse, but rather for a law to prohibit its use and punish the offender. Such mixtures, passing into the stomachs of the people, can only enfeeble and destroy the race.

Fruit Colorings.

Without doubt, the most desirable kind of coloring for extracts, essences and syrups is the juice of natural fruits. To these there can be no possible objection; the only trouble is the inconvenience in procuring them. The time of year in which the fruit is obtainable does not correspond with the opening of the soda water season. This may be overcome in two ways; either by purchasing them already prepared or preparing them yourself when in season.

German black cherry juice may be purchased and is well suited for such purposes. If you prepare the coloring yourself, the black raspberry is, perhaps, the best suited and most convenient. This can be accomplished in the same manner as directed under formulas for the preparation of fruit syrups, except that the amount of sugar may be very much reduced, and the precaution taken to put in dry, well cleaned (scalded), bottles and corked with new and perfect corks, well put in and securely sealed.

In the absence of these colorings, the following formulas may be used:

Solution of Carmine — N. F.

Carmine .... 1 ounce.
Water of Ammonia .... 6 fl. ounces.
Glycerin .... 6 fl. ounces.
Water (enough to make) ....16 fl. ounces.

Triturate the Carmine to a fine powder in a wedgewood mortar, gradually add the Water of Ammonia, and afterward the Glycerin, under constant trituration; transfer the mixture to a porcelain capsule and heat it upon a water bath, constantly stirring until the liquid is entirely free from ammoniacal odor; then cool and add enough water to make sixteen (16) fluidounces.

Solution of Cochineal — N. F.

(COCHINEAL COLOR)

Cochineal (No. 50 powder)	1 troyounce.
Carbonate of Potassium	½ troyounce.
Alum	⅛ troyounce.
Bitartrate of Potassium	1 troyounce.
Glycerin	8 fl. ounces.
Alcohol	1 fl. ounce.
Water (enough to make)	16 fl. ounces.

Triturate the Cochineal intimately with the Carbonate of Potassium and eight (8) fluidounces of water; then add the Alum and Bitartrate of Potassium successively; heat the mixture to boiling, in a capacious vessel, then set aside to cool; add it to the Glycerin and Alcohol, filter and pass enough water through the filter to make sixteen (16) fluidounces.

Tincture of Cochineal — Br.

Cochineal (powdered), 2½ ounces.
Proof Spirit .. .. 1 pint, Imp.=20 fl. oz.

Mix, macerate for seven (7) days in a closed vessel, with occasional agitation; strain, press, filter, and add sufficient Proof Spirit to make one (1) pint (imperial measure).

This color is brightened by acids and deepened by alkalies.

Solution of Cochineal.
(COCHINEAL COLOR)

Cochineal (powdered)	½ ounce.
Alum	½ ounce.
Cream Tartar	½ ounce.
Carbonate of Potash	½ ounce.
Alcohol	½ fl. ounce.
Water	3½ fl. ounces.

Dissolve the Salts in the Water, and when effervescence has ceased add the Cochineal and Alcohol; allow the mixture to stand twenty-four (24) hours, stirring frequently, and filter, adding through the filter enough water to make four (4) fluidounces.

This preparation is liable to spoil by fermentation, and should, therefore, be made in small quantities and kept in a cool place.

Tincture of Cudbear Compound-N.F.

Cudbear	120 grains.
Caramel	1½ troyounces.
Alcohol and Water (of each enough to make)	16 fl. ounces.

Mix Alcohol, one (1) part, and Water, two (2) parts, and macerate Cudbear with twelve

(12) fluidounces for twelve (12) hours; filter through paper and add caramel, previously dissolved in two (2) fluid ounces of water and add sufficient of the above menstruum, through the filter, to make sixteen (16) fluidounces.

Tincture of Saffron — U. S.

Saffron . . . . 3 ounces.
Dilute Alcohol (sufficient to make). .1 pint.

Moisten the Saffron with two and one-half ($2\frac{1}{2}$) fluidounces of the menstruum, and macerate for twenty-four (24) hours; then pack firmly in a cylindrical percolator and gradually pour on the menstruum until two (2) parts of tincture are obtained.

Tincture of Safflower.

(TINCTURE OF AMERICAN SAFFRON)

Safflower . . . . 3 ounces.
Dilute Alcohol (sufficient to make), 1 pint.

Proceed in the same manner as directed for tincture of saffron.

Tincture of Turmeric.

Turmeric (in powder) .... 2½ ounces.
Dilute Alcohol (sufficient to make), 1 pint.

Moisten the Turmeric with a small quantity of the menstruum and pack in percolator; pour on menstruum until one (1) pint of the tincture is obtained.

This is a beautiful yellow liquid, which is rapidly changed to reddish brown by addition of alkalies. Acids produce a light yellow color.

Solution of Caramel.

(CARAMEL BROWN—SUGAR COLOR)

Caramel is manufactured from sugar by carefully applying heat, 400° to 420°; on cooling it is found to be in the form of a black, stiff paste or mass. In Europe it is made from starch sugar (glucose), either by a simple process of heating or heating with addition of carbonate of soda or carbonate of ammonia.

The caramel of the market may be produced by taking—

Sugar (granulated).............. 16 ounces.
Water .... 6 fl. ounces.
Or a sufficient quantity.

Heat the Sugar as above directed, in a porcelain-lined iron vessel, on a sand bath, until entirely liquid and black; while still hot, gradually and carefully pour in the Water, stirring until thoroughly mixed; let stand until partially cooled; then add water enough to make one (1) pint, and strain through purified cotton placed in the neck of a funnel.

PART SIXTH

Appendix.

As to the use of colorings in flavoring extracts, syrups, and so forth, there can be no doubt that those of vegetable or others of harmless origin only, should be recommended and used. We are equally sure that aniline colors are often substituted.

Appendix.

It would appear impossible to attempt to close an effort of this kind without the addition of an appendix. While another score of years may suffice to bring order out of chaos, in the nomenclature of such a work, we feel that an effort such as this must of necessity be very imperfect.

As before stated, we have tried to illuminate the fundamental work as well as the details of formulas, and will now say that, although we think the subject of flavoring extracts and essences is capable of being well defined, in each of their particular spheres, yet the addition of colorings, acidifying solutions, and the various substances usually added in the preparation of syrups (which, so far as the druggist is directly concerned, is the particular aim of such preparations), we would observe that to classify these does appear impractical. So far as colorings go, we have placed them in a "part" by themselves, but as to acid solutions, albumen solutions, and the like, we can only throw them together, as a banker just beginning business would his small change, to be properly arranged when it shall have grown to larger proportions.

Preservation of Lemons.

Mr. George Mee, of London, England, says he has for some time adopted a plan for securing fresh lemons at all seasons of the year, by simply varnishing them with a solution of shellac in spirit of wine. As an experiment, we kept a lemon for many months in this way.

When the peel is desired for flavoring, the lemon may be simply kneaded with the hands. The skin of shellac readily peels off and leaves the rind quite unimpaired. This process, we suppose, will apply to oranges also.

Grating.

In order successfully to grate the rind of lemons and oranges, the grater should be extremely coarse. This is absolutely necessary, owing to the tendency of the moist rind to cake and pack into the openings of the grater. If this is properly looked to, the grating will be found a very easy task, otherwise, a very difficult one.

Restoring Essential Oils.

Essential oils which have become deteriorated in odor by age, may be restored by shaking for fifteen minutes with a thick cream, made of powdered borax, animal charcoal and water, and then filtering. If much discolored, mix with equal weight of poppy oil and a saturated solution of common salt, and distill. The coloring matter is left behind in the fixed oil.

Adulterated Oil of Bitter Almonds.*

Oil of bitter almonds, adulterated with an admixture of nitro-benzole, may be detected by solution of potassa. The liquid is green if nitro-benzole is present, and upon dilution three layers are formed, the lower yellow, the upper green. Over night the green color changes to red.

*Proctor.   *Zeitscher F. Anal. Chem., 1872, p. 316.

Prification of Oil of Bitter Almonds.

Oil of Bitter Almonds .... 2 fl. ounces.
Sulphate of Iron (U. S.) .... 1 ounce.
Lime (recently burnt) .... ½ ounce.

Water, a sufficient quantity.

Dissolve the Sulphate of Iron in one-half (½) pint of Water; slake the Lime with one-half (½) pint of the same fluid, and mix them; shake this thoroughly with the Oil of Bitter Almonds in a bottle, then distill in a glass retort or small tin still, with a good refrigeratory, until the purified oil of bitter almonds has all distilled over, which is known by the distilled water ceasing to be milky and odorous. The heavy oil, after allowing time for its separation and subsidence, is removed for use.

Simple Separating Funnel.

Take a glass tube, four or five inches long and one-fourth of an inch in diameter. Close one end by fusing in the flame of a spirit lamp; three-fourths of an inch from this end file a hole with a rat-tail file, moistened with oil of turpentine, in which a small quantity of camphor has been dissolved.

In the small end of funnel place a cork, through which a hole has been made, that will allow the tube to fit tightly. Insert the closed end of the tube in the cork so that the aperture in the side shall be just below the top of the cork. When the contents of the funnel are to be drawn off, push the tube into the funnel until the hole is just above the cork. When the heaviest liquid has passed out, pull the tube down until the hole is just below the cork; this allows the air to enter, and the liquid remaining in the tube to flow out; then by pushing the tube into the funnel as before, the remaining liquid can be drawn off.

Examination of Vanilla Beans.

"After a series of experiments, I confess my disappointment at the results, as I had for years been led to believe that all that was good in Vanilla existed in the rind. * * * I feel safe in asserting that, while the rind contributes much body to an Extract of Vanilla, the delicacy of flavor lies in the pulp.

"An examination of the pulp samples * * * evinces the fact that a menstruum consisting of cologne spirit, three (3) parts, and water, one (1) part, is the proper solvent.

"This menstruum does not seem to be quite as well adapted to the rind as dilute spirit, but the difference is so slight as to be overcome by the greater value of the pulp flavor."

The experiments referred to above involved maceration, continued for a space of five (5) weeks, thoroughly shaking occasionally.

The proportion of bean to menstruum was one ounce to the pint.

Estimation of Oil Present in Flavoring Extracts.

The depths of low degree in quality or attenuation of strength, we do not pretend to delineate.

Invest ten cents in a small bottle of extract, say of Lemon, and test for proportion of oil present, by taking one hundred (100) parts of the extract, also one hundred (100) parts of different strengths or dilutions of your own make; add definite amounts of water to each, and observe the relative extent or density of milkiness. This, in a crude way, will give you a fair estimate of the amount of oil present in the sample purchased.

This method will apply only to any extract made directly from the oil, as a source of flavoring. Those made from the fruit, seed, leaf or plant direct, cannot be examined in this way. These may be diluted with water in definite amounts, and compared carefully as to odor, with samples purchased, for relative strength.

Some Flavoring Ex'ts of the Market.

We have knowledge of one manufacturer who is engaged in preparing flavoring extracts for a jobbing house, who furnishes Extract of Lemon and Vanilla at the uniform price of thirty-five cents a dozen, in boxes, complete. These, he informed us, are sold by the jobber mostly to peddlers at forty-five cents per dozen, who in turn dispose of them to the country stores at the nominal sum of seventy-five cents per dozen, they selling it for — a good article, of course.

The same manufacturer showed us a sample of the Extract of Vanilla, which he claimed cost him sixty-nine cents a gallon in bulk.

Just what it contained, or rather what it did not contain, you might readily guess. It had color, looked dark; had odor, a smell of something; and the liquid, when applied to the tongue, so far as alcohol was concerned, appeared to be a stranger. "Further this deponent sayeth not."

Soluble Extracts from Volatile Oils.

Various formulas have been recommended for securing the solubility of Extracts of Lemon and Orange, the bases of which, Essential Oils, even when in comparatively dilute solution, do not form with syrup or water perfectly clear mixtures.

It has been suggested in an English journal to shake the Oil of Lemon with five times its volume of alcohol, allowing the mixture to become clear, and then drawing off the spirituous portion. This portion, it is said, contains the flavoring principle, and will mix clear with syrup, and we presume with water.

Carbonate of Magnesia, Powdered Pumice or Purified Talcum may be used for the purpose of making Soluble Extracts.

The following formula will answer for either *Lemon* or *Orange Extract:*

Soluble Extract of Lemon.

Carbonate of Magnesia, or Powdered Pumice, or Purified Talcum	1 ounce.
Oil of Lemon	¾ fl. ounce.
Alcohol (deodorized)	6 fl. ounces.
Water (hot, sufficient to make)	16 fl. ounces.

Dissolve the Oil of Lemon in the Alcohol and mix well with the Carbonate of Magnesia, Powdered Pumice, or Purified Talcum in a mortar; add enough Hot Water to make one (1) pint, and pour into a bottle having twice the necessary capacity; allow to stand for five or six days, shaking often; filter through paper, adding enough water through the filter to make one (1) pint.

"Other formulas, containing more of the Volatile Oil, could be given, with additions of substances having a chemical effect on the oil. This, however, would be expected to impair the flavor." *

---

*"Concentrated Essential Oils, prepared by Hænsel, are said to be from two to thirty times stronger than the ordinary oils from which they are distilled They mix clear with alcohol, diluted to as low as sixty-five (65) per cent.; and medicated waters can be made from them without having recourse to Magnesia, Talcum, and so forth."

Soluble Extract of Ginger. *a*

(SOLUTION OF GINGER. SOLUBLE ESSENCE(?) OF GINGER)

Fluid Extract of Ginger (U. S.)   4 fl. ounces.
Pumice* (in moderately fine
   powder) .... 1 troyounce.
Water (enough to make)........12 fl. ounces.

Pour the Fluid Extract of Ginger into a bottle, add to it the Pumice, and shake the mixture thoroughly and repeat in the course of several hours; then add the Water in proportions of about two (2) fluidounces, shaking well and frequently after each addition; when all is added, repeat the agitation occasionally during twenty-four hours; then filter, returning the first portion of the filtrate until it runs through clear; and, if necessary, pass enough Water through the filter to make twelve (12) fluidounces.

Soluble Extract of Ginger. *b*

(SOLUTION OF GINGER. SOLUBLE ESSENCE(?) OF GINGER)

Jamaica Ginger (ground) .... 2 pounds.
Pumice Stone (powdered)........ 2 ounces.

---
*We would suggest that the Powdered Pumice be repeatedly and carefully washed before being used in the above operation.

Lime (slaked) .... 2 ounces.
Dilute Alcohol (sufficient to make), 4 pints.

Rub the Ginger with the Pumice Stone and Lime thoroughly mixed; then moisten with Dilute Alcohol until perfectly saturated; place the mixture in a narrow percolator, being careful not to use any force in packing; simply place it to obtain the position of a powder required to be percolated, so that the menstruum will go through uniformly; lastly add Dilute Alcohol and proceed until four (4) pints of the percolate are obtained; allow the liquid to stand for twenty-four hours and filter, if necessary.

Jamaica Ginger is to be preferred in making this extract, as it not only has a finer flavor than the commoner kinds but is freer from resin.

Solution of Acid Phosphates.

(PHOSPHATE SOLUTIONS.)

This compound, originally introduced as a proprietary article, is now universally used as a medicine and at the fountain. Two forms, with and without Iron, are in common use.

NOTE — The same remarks as to Powdered Pumice will apply in this formula.

WITHOUT IRON—OR SIMPLE.

Phosphate of Lime .... 384 grains.
Phosphate of Magnesia .... 256 grains.
Phosphate of Potassa .... 192 grains.
Phosphoric Acid (60 per cent.) .... 640 minims.
Water (sufficient to make) .... 1 pint.

WITH IRON—OR COMPOUND.

Phosphate of Lime .... 384 grains.
Phosphate of Magnesia .... 64 grains.
Phosphate of Potassa .... 32 grains.
Phosphate of Iron .... 64 grains.
Phosphoric Acid (60 per cent.) .... 816 minims.
Water (sufficient to make) .... 1 pint.

In either formula, mix the Phosphates and Phosphoric Acid together in a wedgewood or porcelain mortar, add the water, and stir until dissolved. Filter, if necessary.

The sixty per cent. *Phosphoric Acid* may be readily produced from the U. S. Acid, which is fifty per cent., by simply taking one hundred (100) parts of the former and evaporating, in a porcelain capsule, to eighty (80) parts, thereby increasing the per cent. by ten (10). The same result may be attained by adding twenty (20) per cent. more of the official or fifty (50) per cent. acid, allowing this to replace an equal quantity of Water in the formula.

Compound Phosphate Solution.

Magnesium Carbonate .... 600 grains.
Calcium Carbonate .... 600 grains.
Potassium Bicarbonate .... 600 grains.
Phosphoric Acid (U. S.) .... 10 fl. ounces.
Water (sufficient to make) .... 5 pints.

Mix and prepare as directed under previous formula.

One to two teaspoonfuls of this solution is considered a dose. Diluted, of course.

Fruit Acid Solution.
(SOLUTION OF CITRIC ACID)

Citric Acid .... 4 ounces.
Water .... 8 fl. ounces.

Mix and make solution, and filter, if necessary.

One (1) to two (2) fluiddrams or more may be added to a quart of syrups not already tart, if desired to be made so.

Acid Solutions are sometimes prepared from Tartaric Acid, but from its greater liability to deteriorate, it is not recommended, although cheaper than Citric Acid.

This has already been dilated on under syrups, but it would appear also necessary to note the fact here, as being under this particular head and for that reason more likely to be noted and remembered.

Other strengths of this solution are recommended and sold. You can determine the proportion of acid approximately in any Citric Acid Solution by evaporating the solution and weighing.

Soda Fountain "Mixtures."

CALISAYA CORDIAL

Elixir of Calisaya .... 8 fl. ounces.
Orange Syrup .... 24 fl. ounces.
  Mix well.

TONIC HOPOPHOSPHITES.

Syrup Hypophosphite Lime and
  Soda .... 4 fl. ounces.
Vanilla Syrup .... 28 fl. ounces.
  Mix well.

BEEF, WINE AND IRON.

Beef, Wine and Iron .... 8 fl. ounces.
Vanilla Syrup .... 24 fl. ounces.
  Mix well.

COCA TONIC.
Coca Wine .... 8 fl. ounces.
Orange Syrup .... 24 fl. ounces.
Mix well.

COCA AND CALISAYA.
Coca Wine .... 4 fl. ounces.
Calisaya Elixir................. 4 fl. ounces.
Orange Syrup .... 24 fl. ounces.
Mix well.

The above are a few samples of Soda Water "Mixtures" now in use.

Foam.

It has been stated that any harmless mucilaginous substance will answer to hold the foam on soda water. Gum Arabic, Gelatin, the White of Egg, and more recently the Mucilage of Irish Moss, have been recommended and used for such purposes.

Of all the articles used for causing soda water to froth or foam, we find nothing that will answer the purpose to the entire satisfaction of the public taste, half so well as albumen in the form of white of egg.

It is also guaranteed to fill up a drain pipe much more rapidly than any other known substance.

Solution of Albumen.

May be made by adding the White of one (1) Egg to one (1) pint of Water, stirring well, and after standing for a short time, straining; or better, if it is not all to be used at once, adding to one-half (½) pint of Water, straining and adding equal quantity of syrup. The syrup tending to preserve it.

This can be used in the proportion of the White of one Egg to the gallon of Syrup, being one-half greater quantity than is usually recommended.

Solution of Irish Moss.

Irish Moss	1 ounce.
Water (sufficient to make)	1 pint.

Wash the Irish Moss with Water, to free from impurities; add one (1) pint of Water and boil for five (5) minutes or heat in a water bath for fifteen (15) minutes, or macerate in cold water for twenty-four (24) hours, with occasional stirring; filter through purified cotton, on a muslin strainer, in a hot water funnel.

This mucilage, it is claimed, has no more taste than mucilage of Gum Arabic, and is said to keep much better.

It can be used with soda syrups, in the proportion of from two (2) to four (4) ounces of the Irish Moss to one (1) gallon of the syrup.

Solution of Gum Arabic.

Gum Arabic .... 8 ounces.
Water .... 1 pint.

Wash the Gum Arabic with Water, to free it from impurities, and add the Water, stirring occasionally until dissolved.

Used in the proportion of three (3) or four (4) ounces of the Gum Arabic to one (1) gallon of syrup.

Tincture of Quillaia — N. F.

Quillaia (in fine chips) .... 8 troyounces.
Alcohol .... 1 pint.
Water (enough to make)........ 3 pints.

Put the Quillaia in a suitable vessel with two (2) pints of Water, and boil it for fifteen minutes, then strain and add enough Water through the strainer to make the strained decoction, when cold, measure two (2) pints; pour this into a bottle containing the Alcohol; let the mixture stand twelve hours then filter through paper, and add enough Water through the filtrate to make it measure three (3) pints.

Compound Soda Foam.

Sarsaparilla Root (ground fine) .... 8 ounces.
Quillaia Bark (ground fine) .... 8 ounces.
Diluted Alcohol (sufficient to make), 4 pints.

Prepare by percolation.

One (1) to two (2) ounces of this Soda Foam is sufficient for one (1) gallon of Syrup. It produces an excellent foam and is quite tasteless, and is said to be capable of superseding entirely the various substances in common use.

Ales, Beers, Wines, Etc.*

Such an addition to an already well mixed aggregation of formulas, peculiar to the pharmacist, would appear preposterous. In this addition we have two purposes. First, to furnish formulas for a few favorite flavorings for use at the fountain.

*It will scarcely be necessary to add in this connection that by taking advantage of the above suggestions and adding thereto those given under *Soluble Extract of Ginger*, one may readily produce a *Soluble Exteact of Ginger Ale*, etc.

Formerly these drinks were made complete by the formulas given; latterly the first or flavoring parts have been separated and combined with syrup so as to be readily used as are other soda water flavorings.

This may be accomplished by comparing the quantity of the finished preparation, as given herein, and the quantity of syrup it is desired to prepare, with the combined quantity of the flavoring principles in any given formula, and thereby make an extract and syrup for fountain use.

The second purpose is to give a means by which any druggist can furnish his customers the formulas and information necessary for the preparation of the various articles named under this head, without the trouble and time usually required to find them.

GINGER BEER OR ALE. [a]

Ginger (Jamaica, bruised)	5 ounces.
Lemons (sliced)	10 in number.
Cream Tartar	4 ounces.
Sugar (granulated)	5 pounds.
Water (boiling)	5 gallons.

Mix and place in a covered vessel until cool, stirring occasionally; when lukewarm add ten (10) fluidounces of Yeast, and keep in a warm place; after fermenting one day, strain through flannel, let stand a short time, take off scum and bottle.

GINGER BEER OR ALE *b*

Ginger (Jamaica, bruised)	1 ounce.
Lemons (sliced)	1 in number.
Sugar (granulated)	1½ pounds.
Cream Tartar	3 ounces.
Water (boiling)	1½ gallons.
Yeast	1 fl. ounce.

Mix and proceed as directed under Ginger Beer or Ale *a*.

ROOT BEER OR ALE.

Fluid Extract Sarsaparilla (American)	10 fl. drams.
Fluid Extract Pipsissaway	10 fl. drams.
Fluid Extract Wintergreen	4 fl. drams.
Fluid Extract Licorice	4 fl. drams.
Oil Wintergreen	48 minims.
Oil Sassafras	24 minims.
Oil Cloves	12 minims.

Mix and add nine (9) gallons of Water, one (1) gallon of Refined Molasses, and lastly one (1) quart of Yeast, proceeding as directed under Ginger Beer or Ale *a*.

SPRUCE BEER *a*

Essence of Spruce	4 fl. ounces.
Sugar (granulated)	10 pounds.
Water (boiling)	10 gallons.
Yeast	8 fl. ounces.

Mix the above, excepting the Yeast, and proceed as directed under Ginger Beer or Ale *a*.

Various spices may be used for flavoring; also three (3) or four (4) sliced lemons may be added.

SPRUCE BEER *b*

Essence of Spruce*	8 fl. ounces.
Pimenta (bruised),	
Ginger (bruised),	
Hops, (fresh, of each)	4 ounces.
Water	3 gallons.

Mix and boil for five (5) or ten (10) minutes; then strain and add Warm Water eleven

---

NOTE—"The *Abies nigra* (*Pinus nigra*), or *black spruce* of this country yields the '*Essence*' *of Spruce*, which is prepared by boiling the young branches in water and evaporating the decoction. It is a thick liquid, with the color and consistence of molasses, and a bitterish, acidulous, astringent taste."

(11) gallons, Yeast one (1) pint, Molasses six (6) pints; mix well and allow the mixture to stand for twenty-four (24) hours, then bottle or jug for use.

GINGER WINE.

Ginger (Jamaica, bruised)....... 4 ounces.
White of Eggs (well beaten)..... 2 in number.
Sugar (granulated) .... 6 pounds.
Water . ... 16 gallons.

Mix well and boil for fifteen (15) minutes, strain carefully, cool, and add juice of four (4) Lemons, also the rinds for flavoring; yeast eight (8) fluidounces; ferment twenty-four (24) hours in an open vessel in a warm place; put in cask closely bunged, stand two (2) weeks and bottle. A little bushing is said to improve this wine.

MEAD.

(LATIN, HYDROMEL)

Mead is made by fermenting a mixture of Honey and Water, in the usual manner. A little flavor or spice may be added, if desired. Mead is a relic of temperance drinks, handed down to us from the far distant past; common in Rome and, perhaps, obtained by them from Greeks or Egyptians.

Conclusion.

Although this is not a dissertation on Soda Water Syrups or "Mixtures," our task would not be complete until all the information at hand was given. As previously stated, it has to do chiefly with a class of preparations whose ultimate aim, at least so far as the druggist is directly concerned, is combination with syrups and syrup compounds. In these combinations we know no bounds. The genius of man creates daily new forms of flavors and combines remedies for the blood, for the nerves, and for the body in general; also, for the mind or fancy as well. And now we have arrived at the point for which we started, namely, to state that, although no formula for "the same" or "plain" or "you know," nor any of the various alcoholic combinations of endless kinds which, like vice in its various forms, hidden, is not vice to the world, for they do not know it. So these things, hidden by respectability, hurt only those who deal in them.

AMERICAN PHARMACAL JOURNALS.

ESTABLISHED	JOURNAL	ADDRESS	ISSUED	PRICE	
1829	The American Journal of Pharmacy	Philadelphia	Monthly	$3 00	
1857	Druggists' Circular	New York	Monthly	1 50	
1868	Canadian Pharmaceutical Journal	Toronto	Monthly	2 00	
1872	American Druggist	New York	Monthly	1 50	
1878	Western Druggist	Chicago	Monthly	1 00	
1880	Apotheker Zeitung	New York	Semi-Monthly	2 00	
1880	The Druggist	St. Louis	Monthly	1 00	(M. B. & Co.)
1881	Pharmaceutical Record	New York	Weekly	1 50	
1882	Druggists' Journal	Philadelphia	Monthly	1 00	
1882	Indiana Pharmacist	Indianapolis	Monthly	1 00	
1882	National Druggist	St. Louis	Semi-Monthly	1 00	
1882	Pharmaceutische Rundschau	New York	Monthly	2 00	
1883	The Formulary	Westfield, N. Y.	Monthly	1 00	(B. F.)
1884	The Drugman	Nashville	Monthly	1 00	
1885	American Pharmacist	Detroit	Monthly	1 00	(C. W. & Co.)
1887	Registered Pharmacist	Chicago	Monthly	1 00	
1887	Bulletin of Pharmacy	Detroit	Monthly	1 00	(P. D. & Co.)
1887	Omaha Druggist	Omaha	Monthly	1 00	(R. D. & Co.)
1887	Pharmaceutical Era	Detroit	Monthly	1 50	
1887	Drug Topics	New York	Monthly	1 00	(McK. & R.)
1888	Rocky Mountain Druggist	Denver	Monthly	1 00	
1888	New England Druggist	Boston	Monthly	1 50	
1888	Notes on New Remedies	New York	Monthly	1 00	(L. & F.)
1889	Pacific Drug Review	Portland	Monthly	1 00	(S. H. & W.)
1889	Druggists' Gazette	Chicago	Monthly	1 00	
1889	Canadian Druggist	Strathroy, Ont	Monthly	1 00	

INDEX.

Acid Phosphates, Solution of, 141
    Without Iron — Simple... 142
    With Iron — Compound... 142
    Syrup of 114
Acid, Benzoic........................... 86
    Fruit Solution 143
    Oxalic 86
    Phosphoric 142
    Sebacic 86
    Solutions 86
    Succinic 86
    Tartaric 86
Acids, Solutions of, Alcoholic, 86
Adulterated Oil of Bitter
    Almonds 133
Alcohol 22
Alcohol, Atwood's.................... 23
    Deodorized (?)............... 23
    True 23
    Middle Run.................. 23
    Ordinary 22
    Perfumes — Note 23
    True Deodorized............ 23
Alcoholic Solution of Acids.. 86
Albumen, Solution of 146
Ales, Beers, Wines, etc......... 148
Allspice, Extract of 63
Almond, Extract of, *a* 58
    Extract of, *b* 58
    Extract of, *c* 59
Appendix 131
Apple, Essence of................... 81
Apricot, Essence of................ 84
    Syrup of Fruit............... 104
Aromatics, Vegetable........... 27
Articles used in the manufacture of Flavoring Extracts, 21

Artificial Essences................ 73
    Flavors 73
Atwood's Alcohol................. 23

Banana, Essence of................ 84
    Syrup of Fruit............... 105
Beans, Vanilla...................... 33
    Examination of............. 135
    Exhaustion of 36
    Quality of 34
    Tonka......................... 43
Beef, Wine and Iron............. 144
Beer, Ginger, *a*..................... 149
    Ginger, *b* 150
    Root 150
    Root, Syrup of............. 116
    Spruce, *a*.................... 151
    Spruce, *b*.................... 151
Benzoic Acid 86
Bitter Almonds, Extract of..58, 59
Black Cherry, Essence of...... 82
    Pepper, Extract of......... 64

Calisaya Cordial 144
Caramel Brown................... 127
    Solution of.................. 127
Carmine, Solution of—N. F... 123
Capsicum, Extract of........... 64
    Syrup of 115
Celery, Extract of, *a*............ 65
    Extract of, *b* 65
Cherry, Essence of 83
    Black, Essence of......... 82
Chocolate, Extract of 69
    Syrup of, *a*.................. 111
    Syrup of, *b* 111
Cinnamon, Extract of, *a* 61
    Extract of, *b* 62

Citric Acid, Syrup of—U. S.	94
Cloves, Extract of, *a*	63
Extract of, *b*	63
Coca Tonic	145
and Calisaya	145
Cochineal Color	124, 125
Solution of—N. F.	124
Cochineal Coloring, Solution of	125
Tincture of—Br.	124
Coffee, Extract of	70
Syrup of, *a*	112
Syrup of, *b*	112
Syrup of, *c*	113
Syrup of, *d*	113
Cologne Spirit	23
Color, Cochineal	124–125
Colorings	121
Fruit	122
Compound Soda Foam	148
Tincture of Cudbear	125
Cordial, Calisaya	144
Coriander, Extract of	67
Cream Syrup, *a*	107
Syrup, *b*	108
Syrup, *c*	108
Nectar	108
Cudbear, Compound Tincture of	125
Currant, Essence of	85
Deodorized Alcohol (?)	23
True	23
Distilled Water	24
Egg Phosphate, Syrup of	113
Syrup of, one glass	113
Essence of Apricot	84
Apple	81
Banana	84
Black Cherry	82
Cherry	83
Essence of Currant	85
Ginger, Soluble	..
Gooseberry	80
Grape	80
Lemon	82
Melon	79
Orange	81
Peach	84
Pear	82
Pineapple, *a*	75
Pineapple, *b*	76
Plum	83
Raspberry, *a*	78
Raspberry, *b*	79
Strawberry, *a*	76
Strawberry, *b*	76
Strawberry, *c*	77
Essences	73
Artificial	73
Flavoring	73
Fruit	73
Essential Oils	25
Restoring	133
Estimation of Oils Present in Flavoring Extracts	136
Examinat'n of Vanilla Beans	135
Exhaustion of Vanilla Beans	36
Extract of Allspice	63
Bitter Almond, *a*	58
Bitter Almond, *b*	58
Bitter Almond, *c*	59
Black Pepper	64
Celery, *a*	65
Celery, *b*	65
Chocolate	69
Cinnamon, *a*	61
Cinnamon, *b*	62
Cloves, *a*	63
Cloves, *b*	63
Coffee	70
Coriander	67

FLAVORING EXTRACTS. 157

Extract of Capsicum 64
    Ginger 64
        Soluble, *a* 140
        Soluble, *b* 140
    Lemon, *a* 52
    Lemon, *b* 53
    Lemon, Improved 53
    Lemon, Standard 54
    Lemon, Soluble 139
    Mace 62
    Nectarine 61
    Nutmeg or Mace 62
    Nutmegs 62
    Orange, *a* 57
    Orange, *b* 57
    Pineapple (?) 75
    Pot or Soup Herbs, *a* 65
    Pot or Soup Herbs, *b* 66
    Rose, *a* 59
    Rose, *b* 60
    Rose, *c* 60
    Summer Savory 67
    Sweet Basil 67
    Sweet Marjoram 67
    Thyme 66
    Wintergreen or Teaberry, 68
    Vanilla, *a* 39
    Vanilla, *b* 40
    Vanilla, *c* 40
    Vanilla, *d* 41
    Vanilla, *e* 42
    Vanilla *f* 42
    Vanilla with Tonka 44
    Vanilla, Standard 46
    Vanilla without Vanilla .. 47
Extracts, Lemon 50
    Soluble, from Volatile Oils 138

Flavoring Essences 73
Flavoring Extracts 81
    Manufacture of 31
    Sold, Quality of 32

Flavoring Extracts, Some, of
    the Market 137
Flavors, Artificial 73
Foam 145
    Compound Soda 148
Formulas for Fruit Syrups ... 100
Frambois Syrup 107
Fresh Lemon Peel, Tincture
    of 51
Fruit Acid, Solution of 143
    Essences 73
Fruti Fru, Syrup of 109
Fruit Syrup Formulas 100
Fruit Syrups 101
Funnel, Simple Separating ... 134

Ginger Beer or Ale, *a* 149
    Beer or Ale, *b* 150
    Wine 152
Ginger, Extract of 64
    Soluble Essence of 140
    Soluble Extract of, *a* 140
    Soluble Extract of, *b* 140
    Solution of 140
    Syrup of—U. S. 114
    Syrup of, *a* 115
    Syrup of, *b* 115
    Syrup of, Ale 115
Gooseberry, Essence of 80
Grape, Essence of 80
    Syrup of Fruit 106
Grating 132
Gum Arabic, Solution of 147

Hickorynut or Walnut Cream
    Syrup 110
Hypophosphites, Tonic 144

Irish Moss, Solution of 146
Iron, Malt, and Phosphate,
    Syrup of 117

MONOGRAPH ON

Lemon, Essence of 82
   Extract of, *a* 52
   Extract of, *b* 53
   Extract of, Improved 53
   Extract of, Standard 54
   Extract of, Soluble 139
   Extracts 50
   Oil of 49
   Spirit of 51
   Syrup — U. S. 94
   Syrup, *a* 95
   Syrup, *b* 95
   Syrup, *c* 96
   Syrup, *d* 97
   Tincture of — Br. 51
   Tincture of Fresh Peel 51
Lemons, Preservation of 132

Mace, Extract of 62
Manufacture of Flavoring Extracts 31
Mapel Syrup 107
Mead 152
Melon, Essence of 79
Middle Run Alcohol 23
Mixtures, Soda Fountain 144

Nectar Syrup, *a* 106
Nectar Syrup, *b* 106
   Cream Syrup 108
Nectarine, Extract of 61
Nutmegs, Extract of, *a* 62
   Extract of, *b* 62

Oil of Bitter Almonds, Adulterated 133
   Purification of 134
   Lemon 49
   Orange 55
   Present in Flavoring Extracts, Estimation of 136
Oils, Essential 25
   Restoring 133

Orange, Essence of 81
   Extract of, *a* 57
   Extract of, *b* 57
   Oil of 55
   Syrup of — U. S. 97
   Syrup of, *a* 98
   Syrup of, *b* 98
Ordinary Alcohol 22
   Water 24
Orgeat Syrup 109
Oxalic Acid 86

Pear, Essence of 82
Peach, Essence of 84
   Syrup of, Fruit 105
Pepper, Black, Extract of 64
Perfumers' Alcohol 23
Phosphate, Egg, Syrup of 113
   Solution 141
   Solution, Compound 143
Phosphates, Solution of, Acid 141
   Syrup of, Acid 114
Phosphoric Acid 142
Pineapple, Essence of, *a* 75
   Essence of, *b* 76
   Syrup of Fruit 105
Pot or Soup Herbs, Ex of, *a* . 65
   Extract of, *b* 66
Preservation of Lemons 132
Purification of Oil of Bitter Almonds 134

Quality of Flavoring Extracts Sold 32
   Vanilla Beans 34
Quillaia, Tincture of — N. F. . 147

Raspberry, Essence of, *a* 78
   Essence of, *b* 79
   Syrup of — U. S. 101
   Syrup of, *a* 102
   Syrup of, *b* 103
Restoring Essential Oils 133

FLAVORING EXTRACTS.

Root Beer or Ale.................. 150
  Beer, Syrup of 116
Rose, Extract of, *a*............. 59
  Extract of, *b* 60
  Extract of, *c* 60

Safflower, Tincture of 126
Saffron, Tincture of — U. S.... 126
Sarsaparilla, Extract of, *a*. .. 68
  Extract of, *b* 69
  Syrup of, *a*.................. 116
  Syrup of, *b* 116
Sebacic Acid 86
Sherbet Syrup, *a* 106
Sherbet Syrup, *b* 107
Simple Separating Funnel 134
  Syrup 89
  Syrup — U. S. 90
  Syrup, thin.................... 91
Soda Fountain Mixtures........ 144
  Foam, Compound 148
Soluble Essence (?) of Ginger 140
Soluble Extract of Ginger, *a*, 140
  Extract of Ginger, *b*...... 140
  Extract of Lemon 139
  Extracts from Volatile Oils 138
Solution of Acid Phosphates, 141
  with Iron, Compound 142
  without Iron, Simple 142
  Albumen 146
  Caramel 127
  Carmine — N. F. 123
  Cocheneal — N. F. 124, 125
  Fruit Acid 143
  Ginger, *a*.................... 140
  Ginger, *b*.................... 140
  Gum Arabic.................. 147
  Irish Moss.................... 146
Solutions of Acids, Alcoholic, 86
Solutions, Acid 86

Some Flavoring Extracts of the Market.................. 137
Soup or Pot Herbs, Ex. of, *a*.. 65
  Extract of, *b* 66
Spirit of Lemon — U. S......... 51
  Cologne 23
Spruce Beer, *a* 151
Spruce Beer, *b* 151
Strawberry, Essence of, *a* 76
  Essence of, *b*................ 77
  Essence of, *c*................ 77
  Syrup of, *a*.................. 103
  Syrup of, *b*.................. 104
Succinic Acid.................... 86
Sugar Color 127
Summer Savory, Extract of... 67
Sweet Basil, Extract of........ 67
  Marjoram, Extract of...... 67
  Orange Peel, Tincture of — U. S. 56
Syrup 89
Syrup — U. S 90
  Thin 91
  of Acid Phosphates........ 114
  Apricot, Fruit 104
  Banana, Fruit................ 105
  Capsicum 115
  Chocolate, *a*................ 111
  Chocolate, *b*................ 111
  Citric Acid — U. S. 94
  Coffee, *a*.................... 112
  Coffee, *b*.................... 112
  Coffee, *c*.................... 113
  Coffee, *d*.................... 113
  Egg Phosphate.............. 113
  Egg Phosphate, one glass, 113
  Fruti Fru 100
  Ginger — U. S................ 114
  Ginger, *a*.................... 115
  Ginger, *b*.................... 115
  Ginger Ale.................... 115

Syrup, Grape, Fruit	106
Iron, Malt, and Phosphate	117
Lemon — U. S	94
Lemon, *a*	95
Lemon, *b*	95
Lemon, *c*	96
Lemon, *d*	97
Nectar, *a*	106
Nectar, *b*	106
Orange — U. S	97
Orange, *a*	98
Orange *b*	98
Orange *c*	99
Orange, red	99
Peach, Fruit	105
Plum, Fruit	105
Pineapple Fruit	103
Raspberry — U. S	101
Raspberry, Fruit, *a*	102
Raspberry, Fruit, *b*	103
Root Beer	116
Sarsaparilla, *a*	116
Sarsaparilla, *b*	116
Sherbet, *a*	106
Sherbet, *b*	107
Strawberry, Fruit, *a*	103
Strawberry Fruit, *b*	104
Tamarinds, Fruit	105
Vanilla, *a*	91
Vanilla, *b*	92
Vanilla, *c*	93
Walnut or Hickorynut	110
Syrup Chocolate, *a*	111
Chocolate, *b*	111
Coffee, *a*	112
Coffee, *b*	112
Coffee, *c*	113
Coffee, *d*	113
Cream, *a*	107
Cream, *b*	108
Cream, *c*	108
Solutions, Frambois	107
Maple	107
Necter Cream	108
Orgeat	109
Simple	90
Simple, thin	91
Syrups, Fruit	101
Tamarinds, Fruit, Syrup of	105
Tartaric Acid	86
Teaberry or Wintergreen, Extract of	68
Thin Syrup	91
Thyme, Extract of	66
Tincture of Cochineal — Br	124
Cudbear Compound	125
Fresh Lemon Peel	51
Lemon — Br.	51
Quillaia — N. F.	147
Safflower	126
Saffron — U S.	126
Sweet Orange Peel — U. S.	56
Turmeric	127
Vanilla — U.S.	37
Tonic, Coca	145
Tonic Hypophosphates	144
Tonka Beans	43
True Deodorized Alcohol	23
Turmeric, Tincture of	127
Vanilla Beans	33
Examination of	135
Exhaustion of	36
Quality of	34
Extract of, *a*	39
Extract of, *b*	40
Extract of, *c*	40
Extract of, *d*	41
Extract of, *e*	42
Extract of, *f*	42
Extract of, Standard	46
Extract of, with Tonka	44

Vanilla, Extract of, without Vanilla 47	Walnut or Hickorynut Cream Syrup 110
Syrup of, *a* 91	Water 24
Syrup of, *b* 92	Distilled 24
Syrup of, *c* 92	Ordinary 25
Vegetable Aromatics 27	Wine, Ginger 152
Volatile Oils, Soluble Extracts from 138	Wintergreen or Teaberry, Extract of 68

This work, entitled "Monograph on Flavoring Extracts," is published for the purpose of improving the quality of Extracts produced.

We feel it incumbent on us to say that without good materials your efforts will fail, and you will be disappointed.

You cannot make good Extract of Vanilla from Tonka Beans, and you can not produce anything good from the ordinary Grain Alcohol or the general run of so-called Cologne Spirits.

The successful Perfumers and Extract Manufacturers in this country buy the best material, and continue to buy it after they have gotten themselves established.

We have been to some trouble and expense to find out who makes the best Cologne Spirit in the United States, in the judgment of many leading Perfumers in New York and Philadelphia, and, somewhat to our surprise, find the universal testimony to be that C. H. Graves & Sons, 35 Hawkins Street, Boston, Mass., stand in the front rank of popularity. We have bought Graves' XXX Extra French Cologne Spirits, and experimented with it, and must say, we are delighted with the results.

While this Spirit costs more to start with than the ordinary Cologne Spirits, the saving in Essential Oils and Vanilla Bean will offset the difference in cost fourfold, and the finished extracts are a constant pleasure.

JOSEPH HARROP.

NEW YORK

TYLER & FINCH,

IMPORTERS,

54 Cedar St., NEW YORK

OFFER FROM STORE, AND TO ARRIVE, IN QUANTITIES TO SUIT.

VANILLA BEANS — Mexican, every quality and length. Also Cuts, Splits, Granulated, and Powdored with Sugar.

TONKA BEANS — Angostura.

OIL LEMON,
OIL ORANGE, } Last crop. Finest quality.
OTTO ROSE,

We are one of the oldest houses in this line here, and cordially invite correspondence.

For nearly **half a century** we have advertised "any Vanilla selected by us that is not satisfactory on arrival, may be promptly returned at our cost." We make packages of any size for safe shipment by express or mail to any part of the world.

ROWNTREE'S PURE EXTRACT OF COCOA,

For making Chocolate Syrup FOR USE at the SODA FOUNTAIN.

This extract is in the form of a very fine powder, thus making it very soluble. Contains No Cocoa Butter, Sugar, Starch, or Flour.

Put up in 5 lb. Tins. Price 65c per lb.

FORMULA:

Dissolve **One Half Pound** Rowntree's Extract of Cocoa in one pint of Hot Water. Add that solution to five pounds of Granulated Sugar and One Half Gallon Water, and bring the whole to a gentle boil, flavor with a little Vanilla Extract. The above will make One Gallon of the most delicious Chocolate Syrup.

Or, use one-half pound of the Extract with one gallon of Syrup and mix when the Syrup is hot.

The following will show the dealer the actual cost of producing this most delicious Chocolate syrup, and illustrate the *economy* of using

ROWNTREE'S EXTRACT OF COCOA,

ONE-HALF POUND of which produces (owing to its extreme strength) better results than can be obtained from A POUND of any other Chocolate preparation:

½ lb. Rowntree's Extract of Cocoa at 65c per lb.	.33
5 " Granulated Sugar at 5c per lb.	.25
½ gallon Water,	.00
Extract Vanilla to suit taste, about	.07
Actual cost of one Gallon Chocolate Syrup,	65

Serving 2 ozs. Syrup to a glass (really more than is necessary), 1 gallon, 128 ozs., costing 65c, will draw 64 glasses Chocolate Soda, making the actual cost per glass, ready to serve, but one cent.

In using the ROWNTREE EXTRACT OF COCOA you will not offend your customers by drawing a greasy class of Chocolate Soda, which is very obnoxious as well as unsightly, but you can give them, by using these goods, an article that is more than satisfactory. Send an order for a sample 5 lb. tin, and a trial will convince you it is as represented.

E. C. RICH COMPANY, (Limited,)

Sole Importers and United States Agents.

160 Franklin Street, NEW YORK.

201 State Street, BOSTON.

It is said of a noted anti foreign product personage of the past, that, although he would have only carpets and upholstery of American make, still a Turkish rug or two could be found among his possessions. So we, while presuming to give formulas for the manufacture of flavoring extracts, use and recommend a few for which formulas are not given in this book.

JOSEPH HARROP.

To properly compound the various formulas in this book, it is essential that the ingredients should be of the purest and most select character. They will be so if they come from the laboratory of Beach & Clarridge. No specialists surpass Messrs. B. & C. in experience and ability, and the high quality of their preparations is attested by their thousands of patrons among the leading soda water dispensers of the world.

On application, Beach & Clarridge will mail you, free of charge, their latest catalogue, containing the most complete list of soda water flavors, fruit juices, extracts, essences, tinctures, bitters, fruit and essential oils, fruit acids, and vegetable colors, ever published; also their copyrighted formulas for Boston's most popular mixed drinks.

BEACH & CLARRIDGE,
Boston, Mass., U. S. A.

The FOUNDATION of PERFECT SODA.

Langs' Rock Candy Syrup.

STANDARD
FOR
FIFTY-FIVE YEARS.

WM. LANG & SON,
Rock Candy and Rock Candy Syrup,
207 and 209 Bainbridge Street,
PHILADELPHIA.

1836—ESTABLISHED—1836

BRANCH HOUSE:
28 and 30 West Broadway, NEW YORK.

The Artic Soda Water Apparatus. Low Prices. Easy Terms. The TROUBADOUR SIBERIAN-ARTIC. The most perfect of modern Soda Water apparatus. The factured under the Artic and Siberian patents. Send for Illustrated catalogue and price list. Old apparatus taken in part payment. Second hand apparatus in thorough repair at very low prices, and on particularly easy terms.

Address all communications. **JAMES W. TUFTS, 33 Bowker St., Boston, Mass.** CHICAGO, 84 and 86 Jackson Street. BOSTON, 96 Portland St.

Salesrooms: NEW YORK, 10 Warren St., near Broadway and City Hall.

www.ingramcontent.com/pod-product-compliance
Lightning Source LLC
Chambersburg PA
CBHW011340090426
42744CB00014B/1986